ISAAC T

EXAMPLES OF
ANALYTICAL GEOMETRY
OF THREE DIMENSIONS

Elibron Classics
www.elibron.com

WORKS BY I. TODHUNTER, M.A., F.R.S.

Euclid for Colleges and Schools. New Edition.
18mo. cloth. 3s. 6d.

Mensuration for Beginners. With numerous
Examples. New Edition. 18mo. cloth. 2s. 6d.

Algebra for Beginners. With numerous Examples. New Edition. 18mo. cloth. 2s. 6d.

Key to the Algebra for Beginners. New
Edition. Crown 8vo. cloth. 6s. 6d.

Trigonometry for Beginners. With numerous
Examples. New Edition. 18mo. cloth. 2s. 6d.

Key to the Trigonometry for Beginners.
Crown 8vo. cloth. 8s. 6d.

Mechanics for Beginners. With numerous Examples. Fourth Edition. 18mo. cloth. 4s. 6d.

Natural Philosophy for Beginners. With nu-
merous Examples.
Part I. The Properties of Solid and Fluid Bodies. 18mo. 3s. 6d.
Part II. Sound, Light, and Heat. 18mo. 3s. 6d.

Algebra for the use of Colleges and Schools.
With numerous Examples. New Edition. Crown 8vo. cloth. 7s. 6d.

**Key to the Algebra for the use of Colleges
and Schools.** New Edition. Crown 8vo. cloth. 10s. 6d.

A Treatise on the Theory of Equations. Third
Edition. Crown 8vo. cloth. 7s. 6d.

Plane Trigonometry for Colleges and Schools.
With numerous Examples. Seventh Edition. Crown 8vo. cloth. 5s.

**Key to the Plane Trigonometry for Colleges
and Schools.** Crown 8vo. cloth. 10s. 6d.

**A Treatise on Spherical Trigonometry for the
use of Colleges and Schools.** With numerous Examples. Fourth
Edition. Crown 8vo. cloth. 4s. 6d.

EXAMPLES OF

ANALYTICAL GEOMETRY
OF THREE DIMENSIONS.

EXAMPLES OF

ANALYTICAL GEOMETRY

OF THREE DIMENSIONS.

COLLECTED BY

I. TODHUNTER, M.A., F.R.S.,

HONORARY FELLOW OF ST JOHN'S COLLEGE, CAMBRIDGE.

FOURTH EDITION.

MACMILLAN AND CO.

1878.

𝔊ambridge:

PRINTED BY C. J. CLAY, M.A.
AT THE UNIVERSITY PRESS.

A COLLECTION of examples in illustration of Analytical Geometry of Three Dimensions has long been required both by students and teachers, and the present work is published with the view of supplying the want. These examples have been principally obtained from University and College Examination Papers, but many of them are original. The results of the examples are given at the end of the book, together with hints for the solution of some of them.

I. TODHUNTER.

ST JOHN'S COLLEGE,
July 17th, 1858.

EXAMPLES OF ANALYTICAL GEOMETRY OF THREE DIMENSIONS.

I. *The straight line and plane.*

Let OA, OB, OC be three edges of a rectangular parallelepiped which meet at a point O; take O for the origin and the directions of OA, OB, OC for the axes of x, y, z respectively; complete the parallelepiped; let D be the vertex opposite to A, E that opposite to B, F that opposite to C, G that opposite to O. Let $OA = a$, $OB = b$, $OC = c$; and use these data in the following Examples from 1 to 11.

1. Find the equation to the plane passing through D, E, F.

2. Find the equation to the plane passing through G, A, B.

3. Find the equation to the plane passing through G, O, A; also the equation to the plane passing through G, O, B.

4. Find the equations to the straight line OG.

5. Find the equations to the straight lines EB and AD.

6. Find the length of the perpendicular from the origin on the plane in Example 1.

7. Find the length of the perpendicular from C on the plane in Example 2.

8. Find the angle between the planes in Example 3.

9. Find the angle between the straight line in Example 4, and the normal to the plane in Example 1.

10. Find the angle between the straight lines in Example 5.

11.   Find the equations to the straight line which passes through O, and the centre of the face $AEGF$.

12.   Interpret the equation

$$x^2 + y^2 + z^2 = (x \cos \alpha + y \cos \beta + z \cos \gamma)^2,$$

where $\cos^2 \alpha + \cos^2 \beta + \cos^2 \gamma = 1$.

13.   Find the angle between the straight line $\dfrac{x}{2} = \dfrac{y}{\sqrt{3}} = \dfrac{z}{\sqrt{2}}$, and the straight line $\dfrac{x}{\sqrt{3}} = y,\ z = 0$.

14.   Find the equation to the plane which contains the straight line whose equations are

$$Ax + By + Cz = D, \quad A'x + B'y + C'z = D',$$

and the point (α, β, γ).

15.   Find the equation to the plane which passes through the origin and through the line of intersection of the planes

$$Ax + By + Cz = D, \text{ and } A'x + B'y + C'z = D';$$

and determine the condition that it may bisect the angle between them.

16.   Find the equation to the plane which passes through the two parallel straight lines

$$\frac{x-a}{l} = \frac{y-b}{m} = \frac{z-c}{n}; \quad \frac{x-a'}{l} = \frac{y-b'}{m} = \frac{z-c'}{n}.$$

17.   The equation to *one* plane through the origin bisecting the angle between the straight lines through the origin, the direction cosines of which are $l_1, m_1, n_1,$ and $l_2, m_2, n_2,$ and perpendicular to the plane containing them is

$$(l_1 - l_2)x + (m_1 - m_2)y + (n_1 - n_2)z = 0;$$

and the equation to the *other* plane is

$$(l_1 + l_2)x + (m_1 + m_2)y + (n_1 + n_2)z = 0.$$

18. Shew that the equation to a plane which passes through the point (α, β, γ) and cuts off portions a, b from the axes of x and y respectively is

$$\frac{x}{a} + \frac{y}{b} - 1 = \frac{z}{\gamma}\left\{\frac{\alpha}{a} + \frac{\beta}{b} - 1\right\}.$$

19. Find the equation to the plane which contains a given straight line and is perpendicular to a given plane.

20. Shew that if the straight lines

$$\frac{x}{\alpha} = \frac{y}{\beta} = \frac{z}{\gamma}, \quad \frac{x}{a^2\alpha} = \frac{y}{b^2\beta} = \frac{z}{c^2\gamma}, \quad \frac{x}{l} = \frac{y}{m} = \frac{z}{n},$$

lie in one plane, then

$$\frac{l}{\alpha}(b^2 - c^2) + \frac{m}{\beta}(c^2 - a^2) + \frac{n}{\gamma}(a^2 - b^2) = 0.$$

21. Find the length of the perpendicular from the point $(1, -1, 2)$ on the straight line $x = y = 2z$.

22. Find the equation to the plane which passes through the straight lines

$$\frac{x - a}{l} = \frac{y - b}{m} = \frac{z - c}{n},$$

and $\dfrac{x - a}{l'} = \dfrac{y - b}{m'} = \dfrac{z - c}{n'}.$

23. Find the equations to a straight line which passes through the point (a, b, c) and makes a given angle with the plane

$$Ax + By + Cz = 0.$$

24. Find the equation to the plane perpendicular to a given plane, such that their line of intersection shall lie in one of the co-ordinate planes.

25. If the three adjacent edges of a cube be taken for the co-ordinate axes, find the co-ordinates of the points at which

1—2

a plane perpendicular to the diagonal through the origin and bisecting that diagonal will meet the edges.

26. Determine the plane which contains a given straight line, and makes a given angle with a given plane.

27. A straight line makes an angle of 60° with one axis, and an angle of 45° with another; what angle does it make with the third axis?

28. Interpret $x^2 = y^2 = z^2$.

29. Find the condition which must hold in order that the equations

$$x = cy + bz, \quad y = az + cx, \quad z = bx + ay$$

may represent a straight line; and shew that the equations to the straight line then are

$$\frac{x}{\sqrt{(1 - a^2)}} = \frac{y}{\sqrt{(1 - b^2)}} = \frac{z}{\sqrt{(1 - c^2)}}.$$

30. Through the origin and the line of intersection of the planes

$$x \cos \alpha + y \cos \beta + z \cos \gamma - p = 0,$$

and $x \cos \alpha_1 + y \cos \beta_1 + z \cos \gamma_1 - p_1 = 0,$

a plane is drawn; perpendicular to this plane and through its line of intersection with the plane of (x, y) another plane is drawn: find its equation.

31. Find the equation to the plane which passes through a given point and is perpendicular to the line of intersection of two given planes.

32. From any point P are drawn PM, PN perpendicular to the planes of (z, x) and (z, y); if O be the origin, α, β, γ, θ the angles which OP makes with the co-ordinate planes and the plane OMN, then will

$$\operatorname{cosec}^2 \theta = \operatorname{cosec}^2 \alpha + \operatorname{cosec}^2 \beta + \operatorname{cosec}^2 \gamma.$$

33. Apply the equation to a plane

$$x \cos \alpha + y \cos \beta + z \cos \gamma = \delta$$

to demonstrate the following theorem; a triangle is projected on each of three rectangular planes: shew that the sum of the pyramids which have these projections for bases and a common vertex in the plane of the triangle is equal to the pyramid which has the triangle for base and the origin for vertex.

34. Find the equations to the straight line joining the points (a, b, c) and (a', b', c'); and shew that it will pass through the origin if $aa' + bb' + cc' = \rho\rho'$, where ρ and ρ' are the distances of the points respectively from the origin.

35. Express the equations to the straight line

$$a_1 x + b_1 y + c_1 z = a_2 x + b_2 y + c_2 z = a_3 x + b_3 y + c_3 z$$

in the form

$$\frac{x}{l} = \frac{y}{m} = \frac{z}{n}.$$

36. Shew that the equations

$$\frac{x^3 + 1}{x + 1} = \frac{y^3 + 1}{y + 1} = \frac{z^3 + 1}{z + 1}$$

represent four straight lines, and that the angle between any two of them $= \cos^{-1}\left(\dfrac{1}{3}\right)$.

37. The equations to two planes are

$$lx + my + nz = p, \quad l'x + m'y + n'z = p',$$

where

$$l^2 + m^2 + n^2 = 1, \quad l'^2 + m'^2 + n'^2 = 1;$$

find the lengths of the perpendiculars from the origin on the two planes which pass through their line of intersection and bisect the angles between them.

38. Find the equation to a plane parallel to two given straight lines; hence determine the shortest distance between two given straight lines.

39. Find the equation to a plane which passes through two given points and is perpendicular to a given plane.

40. There are n planes of which no two are parallel to each other, no three are parallel to the same straight line, and no four pass through the same point: shew that the number of lines of intersection of the planes is $\frac{n}{2}(n-1)$, and that the number of points of intersection of those lines is

$$\frac{n(n-1)(n-2)}{1 \cdot 2 \cdot 3}.$$

41. Find the shortest distance between the point (α, β, γ) and the plane

$$Ax + By + Cz = D.$$

42. Find the equation to the plane which passes through the origin and makes equal angles with three given straight lines which pass through the origin.

43. Determine the co-ordinates of the point which divides in a given ratio the distance between two points.

Hence shew that the equation

$$Ax + By + Cz = D$$

must represent a *plane*, according to Euclid's definition of a plane.

44. Three planes meet at a point, and through the line of intersection of each pair a plane is drawn perpendicular to the third: shew that in general the planes thus drawn pass through the same straight line.

45. The equations to a straight line are

$$\frac{\alpha - cy + bz}{a} = \frac{\beta - az + cx}{b} = \frac{\gamma - bx + ay}{c};$$

express them in the ordinary symmetrical form.

46. Find the condition which must subsist in order that the equations

$$a + mz - ny = 0, \quad b + nx - lz = 0, \quad c + ly - mx = 0$$

may represent a straight line; and supposing this condition to be satisfied put the equations in the ordinary symmetrical form.

47. Supposing the equations in the preceding Example to represent a straight line, find in a symmetrical form the equations to the straight line from the origin perpendicular to the given straight line; also determine the co-ordinates of the point of intersection.

48. The locus of the middle points of all straight lines parallel to a fixed plane and terminated by two fixed straight lines which do not intersect is a straight line.

49. The equation to a plane is $lx + my + nz = 0$; find the equations to a straight line lying in this plane and bisecting the angle formed by the intersections of the given plane with the co-ordinate planes of (z, x) and (z, y).

50. A straight line, whose equations are given, intersects the co-ordinate planes at three points: find the angles included between the straight lines which join these points with the origin; and if these angles (α, β, γ) be given, shew that the equation to the surface traced out by the straight line in all positions is

$$x \sqrt{(\tan \alpha)} + y \sqrt{(\tan \beta)} + z \sqrt{(\tan \gamma)} = 0.$$

II. *Surfaces of the second order.*

51. If the normal n at any point of an ellipsoid terminated in the plane of (x, y) make angles α, β, γ with the semiaxes a, b, c, and p be the perpendicular from the centre on the tangent plane, then

$$n \cdot p = c^2, \quad \text{and} \quad p^2 = a^2 \cos^2 \alpha + b^2 \cos^2 \beta + c^2 \cos^2 \gamma.$$

52. Find a point on an ellipsoid such that the tangent plane cuts off equal intercepts from the axes. Also find a point such that the intercepts are proportional to the axes.

53. If a, b, c be the semiaxes of an ellipsoid taken in order, and e, e' the excentricities of the principal sections containing the mean axis, shew that the perpendiculars from the centre on the tangent planes at every point of the section of the surface made by the plane $abe'z = c^2ex$ are equal.

54. From a given point O, a straight line OP is drawn meeting a given plane at Q, and the rectangle $OP \cdot OQ$ is invariable: find the locus of P.

55. Sections of an ellipsoid are made by planes which all contain the least axis: find the locus of the foci of the sections.

56. Find the locus of a point which is equidistant from every point of the circle determined by the equations

$$x^2 + y^2 + z^2 = a^2, \qquad lx + my + nz = p.$$

57. Shew that the section of the surface $z^2 = xy$ by the plane $z = x + y + c$ is a circle.

58. Interpret the equation

$$x^2 + y^2 + z^2 = (lx + my + nz)^2.$$

59. A, B, C are three fixed points, and P a point in space such that $PA^2 + PB^2 = PC^2$: find the locus of P, and explain the result when ABC is a right or obtuse angle.

60. Shew that the equation $\frac{x^2}{a^2} + \frac{y^2}{b^2} + \frac{z^2}{c^2} = 1$, where a, b, c are in order of magnitude, may be written thus,

$$k^2 (x^2 + y^2 + z^2 - b^2) - x^2 + m^2 z^2 = 0,$$

where k and m are certain constants. Hence shew that two circular sections of an ellipsoid can be obtained by cutting it by planes passing through its mean axis.

61. A sphere (C) and a plane are given: shew that if any sphere (C') be described touching the plane at a given point and cutting C, the plane of section always contains a given straight line. Shew also that if the point of contact be not given, and if the plane of section always contain a given point, the centre of the sphere C' will always be upon a given paraboloid.

62. If A, B, C be extremities of the axes of an ellipsoid, and AC, BC be the principal sections containing the least axis, find the equations to the two cones whose vertices are A, B, and bases BC, AC respectively; shew that the cones have a common tangent plane, and a common parabolic section, the plane of the parabola and the tangent plane intersecting the ellipsoid in ellipses, the area of one of which is double that of the other; and if l be the latus rectum of the parabola, l_1, l_2 of the sections AC, BC, shew that

$$\frac{1}{l^2} = \frac{1}{l_1^2} + \frac{1}{l_2^2}.$$

63. Tangent planes are drawn to an ellipsoid from a given external point: find the equation to the cone which has its vertex at the origin and passes through all the points of contact of the tangent planes with the ellipsoid.

64. If tangent planes be drawn to an ellipsoid from any point in a plane parallel to that of (x, y), the curve which contains all the points of contact will lie in a plane which always cuts the axis of z at the same point.

65. Shew that the tangent plane to an ellipsoid is expressed by the equation

$$lx + my + nz = \sqrt{(l^2 a^2 + m^2 b^2 + n^2 c^2)}.$$

66. Form the equation to the plane which passes through à given point of an ellipsoid, through the normal at that point, and through the centre of the ellipsoid.

67. A straight line passing through a given point moves so that the projection of any portion of it on a given straight line bears a constant ratio to the length of that portion : find the equation to the surface which it traces out.

68. Find the length of the perpendicular from a given point on a given straight line in space.

Investigate the equation to a right cone, having the axis, vertex, and vertical angle given ; and determine the condition under which the section made by a plane parallel to one of the co-ordinate planes will be an ellipse.

69. Determine the radii of the spheres which touch the co-ordinate planes and the plane $x + y + z = h$.

70. An ellipsoid is intersected in the same curve by a variable sphere, and a variable cylinder ; the cylinder is always parallel to the least axis of the ellipsoid, and the centre of the sphere is always at the focus of a principal section containing this axis. Shew that the axis of the cylinder is invariable in position, and that the area of its transverse section varies as the surface of the sphere.

71. Three edges of a tetrahedron, in length a, b, and c, are mutually at right angles : shew that if these three edges be taken as axes, the equation to the cone which has the origin for vertex, and for its base the circle circumscribed about the opposite face, is

$$\left(\frac{b}{c} + \frac{c}{b}\right) yz + \left(\frac{c}{a} + \frac{a}{c}\right) zx + \left(\frac{a}{b} + \frac{b}{a}\right) xy = 0,$$

and that the plane $ax + by + cz = 0$ is parallel to the sub-contrary sections of the cone.

Find the corresponding equations when either of the other angular points of the tetrahedron is taken as vertex.

72. Tangent planes to the surface whose equation is

$$\frac{x^2}{a^2} + \frac{y^2}{b^2} + \frac{z^2}{c^2} = 1$$

pass through a point P: shew that a sphere can be described through the curve of contact provided P be on a certain straight line passing through the origin.

73. A sphere touches each of two straight lines which are inclined to each other at a right angle but do not meet: shew that the locus of its centre is an hyperbolic paraboloid.

74. Shew that by properly choosing the rectangular axes any two straight lines may be represented by the equations

$$\begin{aligned} y = mx \\ z = c \end{aligned} \Bigg\}, \qquad \begin{aligned} y = -mx \\ z = -c \end{aligned} \Bigg\}.$$

Determine the locus of a point which moves so as always to be equally distant from two given straight lines.

75. A tangent plane to an ellipsoid includes between itself and the co-ordinate planes a constant volume: find the locus of the points of contact.

76. A tangent plane to an ellipsoid passes through a fixed point in one of the axes produced: find the point of contact when the volume between this plane and the co-ordinate planes is a minimum.

77. Prove that the locus of a point whose distance from a fixed point is always in a given ratio to its distance from a fixed straight line is a surface of revolution of the second degree.

78. Two systems of rectangular axes have the same origin: if a plane, or an ellipsoid whose centre is the origin, cut them at distances a, b, c, or a', b', c', respectively, then

$$\frac{1}{a^2} + \frac{1}{b^2} + \frac{1}{c^2} = \frac{1}{a'^2} + \frac{1}{b'^2} + \frac{1}{c'^2}.$$

79. If a moveable straight line make with any number of fixed straight lines the angles θ_1, θ_2, θ_3, ... so that

$$a_1 \cos \theta_1 + a_2 \cos \theta_2 + a_3 \cos \theta_3 + \ldots = \text{a constant},$$

the straight lines all passing through a fixed point, and a_1, a_2, a_3, ... being constants; shew that the moveable straight line will always lie on the surface of a right cone.

80. Two prolate spheroids have a common focus. If their surfaces meet, the points of intersection will lie in a plane which passes through the line common to the directrix planes.

81. A cone passes through the principal section (b, c) of an ellipsoid: shew that the other section is a curve in a plane perpendicular to that of (b, c) if the vertex of the cone be any point of the surface

$$\frac{z^2}{c^2} + \frac{y^2}{b^2} - \frac{x^2}{a^2} = 1.$$

82. An ellipsoid is cut by a plane parallel to one of its principal planes: shew that all normals to the ellipsoid at points in the curve of section pass through two straight lines situated in the other principal planes.

83. An ellipsoid is constantly touched at two points by a sphere of given radius whose centre moves in one of its principal planes: find the locus of the centre of the sphere, and shew that it will describe a portion of an equilateral hyperbola, if the plane in which it moves be that which contains the greatest and the least axes, and if the foci of the other principal sections be equidistant from the centre of the ellipsoid.

84. An elliptic cylinder is cut by a plane passing through the straight line which touches the base at the extremity of its axis minor : shew that, if the sine of the inclination of the cutting plane to the base do not exceed the excentricity of the base, the locus of the foci of the section will be a circle.

85. Find the curves of intersection of the surface

$$xy + yz + zx = a^2$$

with planes parallel to $x + y + z = 0$.

86. If a sphere be placed in a paraboloid of revolution, shew that the section of the paraboloid made by any plane drawn touching the sphere is a conic section having the point of contact for a focus.

87. Shew that any plane parallel to the plane of (x, z) or (y, z) will meet the surface whose equation is $xy = az$ in a straight line; if a plane be made to pass through that straight line and touch the surface will the plane touch the surface at every point of the straight line?

88. Through any point of the curve of intersection of the surfaces

$$\frac{x^2}{a^2} + \frac{y^2}{b^2} - \frac{z^2}{c^2} = 1, \quad x^2 + y^2 + z^2 = a^2 + b^2 - c^2,$$

two straight lines can be drawn on the former surface at right angles to one another.

89. Prove that the plane $z = mx + ny$ will cut the surface

$$\frac{x^2}{a^2} + \frac{y^2}{b^2} = \frac{z^2}{c^2}$$

in two straight lines which are at right angles to one another if

$$\frac{1}{b^2} + \frac{1}{a^2} + m^2\left(\frac{1}{b^2} - \frac{1}{c^2}\right) + n^2\left(\frac{1}{a^2} - \frac{1}{c^2}\right) = 0.$$

90. Shew that an infinite number of straight lines may be drawn lying on the surface whose equation is

$$\frac{x^2}{a^2} - \frac{y^2}{b^2} = \frac{4z}{c}.$$

If two such straight lines be drawn through a point in the plane of (x, z) whose co-ordinates are (x', z') the angle between them is

$$\cos^{-1}\frac{a^2 - b^2 + cz'}{a^2 + b^2 + cz'}.$$

91. Prove that the points on the surface $\frac{x^2}{a^2} - \frac{y^2}{b^2} = \frac{4z}{c}$, the straight lines through which coincident with the surface are at right angles to each other, lie in a plane parallel to the plane of (x, y) and at a distance from it equal to $\frac{b^2 - a^2}{c}$.

92. Determine the condition necessary in order that the plane $lx + my + nz = 0$ may cut the cone $ayz + bzx + cxy = 0$ in two straight lines at right angles to each other; and shew that in that case the straight line through the origin perpendicular to the plane will also lie on the cone.

93. Find the relations between the coefficients of the equation

$$ax^2 + by^2 + cz^2 + 2a'yz + 2b'xz + 2c'yx + 2a''x + 2b''y + 2c''z = f,$$

that it may represent a surface of revolution.

If $\dfrac{b'c'}{a'} - a = 0,\quad \dfrac{c'a'}{b'} - b = 0,\quad \dfrac{a'b'}{c'} - c = 0,$

what is the nature of the surface?

94. Determine the conditions that the equation

$$ax^2 + by^2 + cz^2 + 2a'yz + 2b'zx + 2c'xy = 0$$

may represent a right circular cone; obtaining them in the form

$$\dfrac{a'b'}{c'} + \dfrac{a'^2 - b'^2}{a - b} = 0,\quad \dfrac{c'a'}{b'} + \dfrac{c'^2 - a'^2}{c - a} = 0.$$

95. Find the relations between the coefficients of the equation

$$ax^2 + by^2 + cz^2 + 2b'xz + 2c'yx + 2a''x + 2b''y + 2c''z = f,$$

that it may represent a surface of revolution.

96. Find the locus of the vertices of all right cones which have the same given ellipse as base.

97. Find the axes of the section of the ellipsoid

$$\dfrac{x^2}{a^2} + \dfrac{y^2}{b^2} + \dfrac{z^2}{c^2} = 1$$

made· by the plane $lx + my + nz = 0$.

98. The semiaxes of an ellipsoid are a, b, c; if planes are drawn through the centre to cut it in sections which have the constant area $\dfrac{\pi abc}{f}$, shew that the normals to these planes will all lie upon the cone

$$x^2(a^2 - f^2) + y^2(b^2 - f^2) + z^2(c^2 - f^2) = 0.$$

99. On every central radius vector of the ellipsoid

$$\frac{x^2}{a^2} + \frac{y^2}{b^2} + \frac{z^2}{c^2} = 1,$$

whose centre is C, is taken a point P such that $\pi k CP$ is the area of the section of the ellipsoid made by the plane through C perpendicular to CP, where k is a straight line of constant length: shew that the equation to the locus of P is

$$a^2x^2 + b^2y^2 + c^2z^2 = \frac{a^2b^2c^2}{k^2}.$$

100. The sum of the squares of the reciprocals of the areas of the sections of an ellipsoid made by any three diametral planes at right angles to each other is constant.

101. An ellipsoid and hyperboloid are concentric and their principal sections are confocal: if a tangent plane be drawn to the asymptotic cone of the hyperboloid, the section of the ellipsoid will be of constant area.

102. A sphere and an ellipsoid which intersect are described about the same point as centre: shew that the product of the areas of the greatest and least sections of the ellipsoid, made by planes passing through the centre and any point of the line of intersection of the two surfaces, will be constant.

103. Find the area of the section of the ellipsoid

$$\frac{x^2}{a^2} + \frac{y^2}{b^2} + \frac{z^2}{c^2} = 1,$$

made by the plane $lx + my + nz = \delta$.

104. Shew that the sections of a surface of the second order made by parallel planes are similar curves. Having given the area of the section of an ellipsoid made by the plane $lx + my + nz = 0$, find the area of the section made by the plane $lx + my + nz = \delta$.

105. If S be the area of a section of an ellipsoid made by a plane at the distance h from the centre, S' that of the parallel section through the centre, and p the perpendicular from the centre on the parallel tangent plane, shew that

$$S = S' \left(1 - \frac{h^2}{p^2}\right).$$

106. Tangent planes are drawn to an ellipsoid from a given point: shew that an ellipsoid similar to the given ellipsoid and similarly situated can be made to pass through the given point, the points of contact, and the centre of the given ellipsoid.

107. Normals are drawn to the ellipsoid

$$\frac{x^2}{a^2} + \frac{y^2}{b^2} + \frac{z^2}{c^2} = 1$$

at the points where it is intersected by the plane $z = h$. Shew that the locus of the intersection of these normals with the plane of (x, y) is the ellipse

$$\frac{x^2 a^2}{(a^2 - c^2)^2} + \frac{y^2 b^2}{(b^2 - c^2)^2} = 1 - \frac{h^2}{c^2}.$$

108. All the normals to the ellipsoid in Example 107 meet the plane of (x, y) within an ellipse whose equation is

$$\frac{a^2 x^2}{(a^2 - c^2)^2} + \frac{b^2 y^2}{(b^2 - c^2)^2} = 1.$$

Determine what is represented by the following equations, from 109 to 116 inclusive.

109. $5y^2 - 2x^2 - z^2 + 4xy - 6yz + 8xz - 1 = 0.$

110. $2y^2 - 5x^2 + 2z^2 + 10xy + 4yz + 4y + 16z + 18 = 0$.

111. $4y^2 - 9x^2 + 2xy + 36x - 8y - 4z - 32 = 0$.

112. $y^2 - 4xy + 4x^2 - 6x + 3z = 0$.

113. $x^2 - y^2 + z^2 - 4xy + 6xz - 2yz = f$.

114. $(ay - bx)^2 + (cx - az)^2 + (bz - cy)^2 = f$.

115. $xy + xz + yz - 2x + 6y - 8z = f$.

116. $xy + yz + xz = a^2$.

117. Find the centre of the surface
$$(\alpha y + \beta x - c\gamma)^2 = \alpha\beta(xy - z^2).$$

118. Find the centre of the surface
$$x^2 + y^2 + z^2 + 4xy - 2xz - 4yz + 2x + 4y - 2z = 0.$$

119. Find the centre of the surface
$$x^2 + 2y^2 + 3z^2 + 2(xy + yz + xz) + x + y + z = 1.$$

120. Shew that the equation $yz + zx + xy = a^2$ may be reduced to
$$2x^2 - y^2 - z^2 = 2a^2$$
by transforming the axes.

121. Shew that the equation
$$x^2 + y^2 + z^2 + xy + yz + zx = a^2$$
represents an oblate spheroid whose polar axis is to its equatorial in the ratio of 1 to 2, and the equations to whose axis are $x = y = z$.

122. Determine what is represented by the equation
$$x^2 + y^2 + z^2 + k(xy + yz + zx) = f.$$

123. If two concentric surfaces of the second order have the same foci for their principal sections they will cut one another everywhere at right angles.

T. A. G.   2

124. Find the locus of the intersection of three planes at right angles to each other, each of which touches one of the following three ellipsoids,

$$\frac{x^2}{a^2}+\frac{y^2}{b^2}+\frac{z^2}{c^2}=1, \quad \frac{x^2}{a^2+h^2}+\frac{y^2}{b^2+h^2}+\frac{z^2}{c^2+h^2}=1,$$

$$\frac{x^2}{a^2+k^2}+\frac{y^2}{b^2+k^2}+\frac{z^2}{c^2+k^2}=1.$$

125. Determine the position of the circular sections of an hyperboloid of two sheets, and shew that the same plane will cut the asymptotic cone in a circle.

126. If x_1, y_1, z_1; x_2, y_2, z_2; x_3, y_3, z_3 be the co-ordinates of the extremities of a set of conjugate diameters of an ellipsoid, shew that

$$x_1^2+x_2^2+x_3^2=a^2, \quad y_1^2+y_2^2+y_3^2=b^2, \quad z_1^2+z_2^2+z_3^2=c^2,$$

$$x_1y_1+x_2y_2+x_3y_3=0, \quad x_1z_1+x_2z_2+x_3z_3=0, \quad y_1z_1+y_2z_2+y_3z_3=0.$$

127. If spheres be described on three semi-conjugate diameters of an ellipsoid as diameters, the locus of their intersection is the surface determined by

$$a^2x^2+b^2y^2+c^2z^2=3(x^2+y^2+z^2)^2.$$

128. A plane is drawn through the extremities of three semi-conjugate diameters of an ellipsoid: find the locus of the intersection of this plane with the perpendicular on it from the centre.

129. Tangent planes at the extremities of three conjugate diameters of an ellipsoid intersect in the ellipsoid whose equation is

$$\frac{x^2}{a^2}+\frac{y^2}{b^2}+\frac{z^2}{c^2}=3.$$

130. A prolate spheroid is cut by any plane through one of its foci: shew that the focus is a focus of the section.

131. Shew that the locus of the diameters of the ellipsoid

$$\frac{x^2}{a^2} + \frac{y^2}{b^2} + \frac{z^2}{c^2} = 1,$$

which are parallel to the chords bisected by tangent planes to the cone

$$\frac{x^2}{a^2} + \frac{y^2}{\beta^2} - \frac{z^2}{\gamma^2} = 0 \text{ is the cone } \frac{a^2 x^2}{a^4} + \frac{\beta^2 y^2}{b^4} - \frac{\gamma^2 z^2}{c^4} = 0.$$

132. If three straight lines at right angles to each other touch the ellipsoid

$$\frac{x^2}{a^2} + \frac{y^2}{b^2} + \frac{z^2}{c^2} = 1,$$

and intersect each other at the point (x', y', z') shew that

$$x'^2(b^2 + c^2) + y'^2(c^2 + a^2) + z'^2(a^2 + b^2) = b^2 c^2 + c^2 a^2 + a^2 b^2.$$

133. Find the greatest angle between the normal at any point of an ellipsoid, and the central radius vector at that point.

134. If four similar and similarly situated surfaces of the second order intersect each other, the planes of their intersections two and two all pass through one point.

135. If three chords be drawn mutually at right angles through a fixed point within a surface of the second order whose equation is $u = 0$, shew that $\Sigma \dfrac{1}{R \cdot r}$ will be constant, where R and r are the two portions into which any one of the chords drawn through the fixed point is divided by that point.

Shew also that the same will be true if instead of the fixed point there be substituted any point on the surface whose equation is $u = c$.

136. Let $\dfrac{x}{l} = \dfrac{y}{m} = \dfrac{z}{n}$ be the equations to a straight line: find the equation to a surface every point of which is at the same distance from this straight line as from the point (a, β, γ); and shew that the plane $lx + my + nz = \delta$ cuts the surface in a straight line.

137. A and B are two similar and concentric ellipsoids, the homologous axes being in the same straight line ; C is a third ellipsoid similar to either of the former, its centre being on the surface of B, and axes parallel to those of A or B: shew that the plane of intersection of A and C is parallel to the tangent plane to B at the centre of C.

138. If a parallelepiped be inscribed in an ellipsoid its edges will be parallel to a system of conjugate diameters.

139. The edges of a parallelepiped are $2a$, $2b$, $2c$: shew that an ellipsoid concentric with it and whose semidiameters parallel to the edges are $a\sqrt{2}$, $b\sqrt{2}$, $c\sqrt{2}$, intersects the faces in ellipses which touch each other and the edges.

140. Two similar and similarly situated ellipsoids are cut by a series of ellipsoids similar and similarly situated to the two given ones, so that the planes of intersection of any one of the series with each of the given ellipsoids make a right angle with one another. Shew that the centres of the series of ellipsoids lie on another ellipsoid.

141. If pyramids be formed between three conjugate diametral planes of an ellipsoid and a tangent plane, so that the products of the intercepted portions of the three conjugate diameters may be the least possible, the volumes of all these pyramids will be equal.

142. If from any point in an ellipsoid three straight lines are drawn mutually at right angles, shew that the plane which passes through their intersections with the surface passes through a point which is fixed so long as the original point is fixed. And shew that if the position of the original point on the surface is changed the locus of the point is an ellipsoid whose semiaxes are

$$\frac{a^2 - 2k^2}{a}, \quad \frac{b^2 - 2k^2}{b}, \quad \frac{c^2 - 2k^2}{c},$$

where

$$\frac{1}{k^2} = \frac{1}{a^2} + \frac{1}{b^2} + \frac{1}{c^2}.$$

143. If two right circular cones whose axes are parallel intersect, the projection of their curve of intersection on the plane passing through the axes is a parabola.

144. An ellipsoid is cut by a plane the distance of which from the centre bears a constant ratio to its distance from the parallel tangent plane : shew that the volume of the cone whose base is the section and vertex the centre of the ellipsoid is invariable.

145. Shew that by properly choosing oblique axes the equation to an ellipsoid may be put in the form

$$x^2 + y^2 + z^2 = a'^2;$$

shew that an infinite number of systems of suitable axes can be found.

146. Determine whether the equation to an hyperboloid of one sheet can always be put in the form

$$x^2 + y^2 - z^2 = a'^2,$$

by properly choosing oblique axes.

147. A straight line revolving round an axis which it does not meet generates an hyperboloid.

148. Shew that the equations to the principal axes of the surface

$$ax^2 + by^2 + cz^2 + 2a'yz + 2b'xz + 2c'xy = 1$$

are

$$x\{a'(\mu+a) - b'c'\} = y\{b'(\mu+b) - c'a'\} = z\{c'(\mu+c) - a'b'\};$$

μ being determined by the equation

$$(\mu+a)(\mu+b)(\mu+c) - a'^2(\mu+a) - b'^2(\mu+b)$$
$$- c'^2(\mu+c) + 2a'b'c' = 0.$$

149. In an ellipsoid the sum of the lengths of three conjugate diameters is greatest when the diameters are equal.

150. A surface of the second order circumscribes a tetrahedron and each face is parallel to the tangent plane at the opposite angular point: shew that the equation to the surface, taking as oblique axes the three edges of the tetrahedron which meet at one of its angular points, is

$$\frac{x^2 - ax}{a^2} + \frac{y^2 - by}{b^2} + \frac{z^2 - cz}{c^2} + \frac{1}{abc}\{ayz + bzx + cxy\} = 0.$$

151. Shew that the centre of the surface in the preceding Example coincides with the centre of gravity of the tetrahedron.

152. If two hyperboloids of one sheet have their corresponding axes equal and parallel, shew that they intersect in a plane curve, the plane of which is parallel to the diametral planes conjugate to the straight line joining their centres; and find where it cuts this straight line.

153. Shew that any plane which contains two parallel generating lines of an hyperboloid of one sheet passes through the centre of the surface.

154. The points on an ellipsoid, the normals at which intersect the normal at a fixed point, lie on a cone of the second order whose vertex is the fixed point.

155. Normals are drawn to an hyperboloid of one sheet at points lying along a generating line: shew that these normals are all parallel to a fixed plane.

156. If one of the co-ordinates of an ellipsoid be produced so that the part produced equals the sum of the other two, its extremity will trace out a concentric ellipsoid whose volume is equal to that of the original ellipsoid.

157. Investigate the conditions necessary in order that the equation

$$ax^2 + by^2 + cz^2 + 2a'yz + 2b'zx + 2c'xy = 0$$

may represent two planes.

158. The equation to a surface of the second order is
$$ax^2 + by^2 + cz^2 + 2a'yz + 2b'zx + 2c'xy$$
$$+ 2a''x + 2b''y + 2c''z + f = 0;$$
shew that the diametral planes will be parallel to a given straight line if
$$aa'^2 + bb'^2 + cc'^2 - abc - 2a'b'c' = 0,$$
and to a given plane if
$$a'^2 - bc = 0, \quad b'^2 - ac = 0, \quad c'^2 - ab = 0.$$

159. Find the straight line in which the two surfaces intersect whose equations are
$$x^2 \cos 2\theta_1 + y^2 \cos 2\theta_2 + z^2 \cos 2\theta_3 + 2xy \cos (\theta_1 + \theta_2)$$
$$+ 2zx \cos (\theta_1 + \theta_3) + 2zy \cos (\theta_2 + \theta_3) = 0,$$
$$x^2 \sin 2\theta_1 + y^2 \sin 2\theta_2 + z^2 \sin 2\theta_3 + 2xy \sin (\theta_1 + \theta_2)$$
$$+ 2zx \sin (\theta_1 + \theta_2) + 2zy \sin (\theta_2 + \theta_3) = 0.$$

160. Shew that the cones determined by
$$\frac{x(yb + zc - xa)}{A} = \frac{y(zc + xa - yb)}{B},$$
and
$$\frac{y(zc + xa - by)}{B} = \frac{z(xa + yb - zc)}{C},$$
intersect in the straight line determined by
$$\frac{x}{A(Bb + Cc - Aa)} = \frac{y}{B(Cc + Aa - Bb)} = \frac{z}{C(Aa + Bb - Cc)};$$
and find the other straight lines in which they intersect.

161. Find the eccentricity of any section of a paraboloid of revolution in terms of the angle of inclination of the cutting plane to the axis.

162. If through any fixed point chords be drawn to an ellipsoid, the intersection of pairs of tangent planes at their extremities all lie in one plane.

163. Let S be a fixed point, and EF, LM two fixed straight lines; take a point P such that if PQ be drawn perpendicular to EF to meet LM at Q, then PQ may bear an invariable ratio to SP; shew that the locus of P will be a surface of the second order. Shew also that when SP, PQ are unequal, a section of the surface by a plane perpendicular to EF is a circle, and that when SP, PQ are equal the surface has in general no centre.

164. The equation $\dfrac{x^2 + y^2}{a^2} + \dfrac{z^2}{c^2} = 1$, by varying c, represents a series of spheroidal surfaces. If normals be drawn to the spheroids from a fixed point in the axis of z, their intersections with the surfaces will lie on the surface of a given sphere.

165. An ellipsoid and hyperboloid of two sheets have the same principal diameters, both in magnitude and position. Let any tangent plane be drawn to the ellipsoid at a point P; this will intersect in general the hyperboloid in a plane curve, through which if there be drawn any three tangent planes to the latter surface they will meet at a point P' which lies on the ellipsoid. Also if P' be the point of contact of a second tangent plane to the ellipsoid, P will be the intersection of a second set of tangent planes to the hyperboloid, drawn as before.

166. The surface of an ellipsoid is represented by the equations

$$x = a \cos \alpha, \qquad y = b \cos \beta, \qquad z = c \cos \gamma;$$

shew that the three straight lines whose direction angles are $(\alpha, \beta, \gamma, \&c.)$ corresponding to any system of conjugate diameters are perpendicular to each other.

If CP_1, CP_2, CP_3 be drawn perpendicular to three conjugate tangent planes

$$\frac{1}{CP_1^2} + \frac{1}{CP_2^2} + \frac{1}{CP_3^2} = \frac{1}{a^2} + \frac{1}{b^2} + \frac{1}{c^2};$$

and if they cut the ellipsoid at points Q_1, Q_2, Q_3, respectively,

$$\frac{1}{CP_1^2 CQ_1^2} + \frac{1}{CP_2^2 CQ_2^2} + \frac{1}{CP_3^2 CQ_3^2} = \frac{1}{a^4} + \frac{1}{b^4} + \frac{1}{c^4}.$$

167. A plane is drawn according to a certain assigned law cutting an ellipsoid; find the locus of the vertex of the cone which touches the ellipsoid in the curve of intersection.

If a, b, c be the semiaxes of the ellipsoid, $\dfrac{x}{\alpha} + \dfrac{y}{\beta} + \dfrac{z}{\gamma} = 1$ the equation to the cutting plane, α, β, γ, being connected by the relation $\dfrac{a^4}{\alpha^2} + \dfrac{b^4}{\beta^2} + \dfrac{c^4}{\gamma^2} = $ constant, the locus is a sphere.

168. Shew that the tangent lines to an ellipsoid from an external point touch it in a plane curve, and find the equations to the curve.

If the external point lies on a fixed plane, shew that the plane of contact passes through a fixed point; determine its co-ordinates.

If the external point lies on a fixed straight line, shew that the plane of contact contains a fixed straight line; determine its equation.

169. If straight lines be drawn from any point of the surface $\dfrac{x^2}{a^4} + \dfrac{y^2}{b^4} + \dfrac{z^2}{c^4} - \dfrac{1}{k^2} = 0$, to touch the surface $\dfrac{x^2}{a^2} + \dfrac{y^2}{b^2} + \dfrac{z^2}{c^2} - 1 = 0$; shew that the planes in which the curves of contact lie all touch the sphere $x^2 + y^2 + z^2 = k^2$.

170. Shew that the right circular cylinders described about the ellipsoid
$$\frac{x^2}{a^2} + \frac{y^2}{b^2} + \frac{z^2}{c^2} = 1,$$
$2b$ being the mean axis, are represented by the equation
$$(b^2 - c^2)\,x^2 - (c^2 - a^2)\,y^2 + (a^2 - b^2)\,z^2$$
$$\pm 2\sqrt{(a^2 - b^2)}\sqrt{(b^2 - c^2)}\,xz = (a^2 - c^2)\,b^2.$$

171. Three cylinders circumscribe a given ellipsoid; the axes of the cylinders are mutually at right angles; shew that the sum of the squares of the areas of sections of the cylinders by planes respectively perpendicular to their axes is constant.

172. If straight lines be drawn from the centre of an ellipsoid (whose semi-axes are a, b, c) parallel to the generating lines of an enveloping cone, the conical surface so formed will intersect the ellipsoid in two planes parallel to the plane of contact. The locus of the vertex of the enveloping cone which causes one of the planes to coincide with the plane of contact is

$$\frac{x^2}{a^2} + \frac{y^2}{b^2} + \frac{z^2}{c^2} = 2.$$

173. Of all cones which envelope an ellipsoid, have their bases in the tangent plane at a given point P, and are of the same altitude, that is the least which has its vertex in the diameter through P; and of all which have their vertices in this diameter, that is the least whose axis is twice that diameter.

174. If a globe be placed upon a table the breadth of the elliptic shadow cast by a candle (considered as a luminous point) will be independent of the position of the globe.

175. In the preceding Example, if an ellipsoid having its least axis vertical be substituted for the globe, determine the condition of the shadow of the globe being circular. It may be shewn that the locus of the luminous point must be an hyperbola, and that the radius of the circular shadow is independent of the mean axis of the ellipsoid.

176. Of a series of cones enveloping an ellipsoid, the vertices lie on a concentric ellipsoid, similar to the given one and similarly situated. Shew that any two cones of the series intersect one another in two planes.

177. Shew that if, in the preceding Example, the vertices are supposed to lie also on a third ellipsoid concentric with the other two and similarly situated, and whose axes are respectively as the squares of theirs, these two planes are at right angles to one another.

178. A cylinder is circumscribed about an ellipsoid, and at the extremities of the diameter parallel to the generating lines of the cylinder tangent planes are drawn: shew that the volume of all cylinders so shut in is constant.

179. The locus of the vertex of a cone which envelopes a given ellipsoid (A) is a straight line passing through the centre of (A); an ellipsoid similar to A and similarly placed has the vertex of the cone for centre and cuts the cone in a curve (B). If the major axis of this ellipsoid vary as the distance of its centre from that of A, shew that the locus of B is an elliptic cylinder.

180. Suppose a cylinder to envelope an ellipsoid, and suppose a tangent plane to be drawn to the ellipsoid at one extremity of the diameter which is parallel to the axis of the cylinder. Let a straight line be drawn from the centre of the ellipsoid to meet the ellipsoid, the above tangent plane, and the enveloping cylinder; and suppose r, s, t to denote the respective distances of the points of intersection from the centre of the ellipsoid. Shew that

$$\frac{1}{r^2} = \frac{1}{s^2} + \frac{1}{t^2}.$$

181. Suppose a cone to envelope an ellipsoid; let R' be the distance of the vertex from the centre of the ellipsoid, $2R$ the length of the diameter of the ellipsoid which is in the direction of the straight line joining the centre of the ellipsoid with the vertex of the cone; and suppose a tangent plane to be drawn to the ellipsoid at one extremity of this diameter. Let a straight line be drawn from the centre of the ellipsoid to meet the ellipsoid, the above tangent plane, and the enveloping cone; and suppose r, s, t to denote the respective distances of the points of intersection from the centre of the ellipsoid. Shew that

$$\left(\frac{1}{Rt} - \frac{1}{R's}\right)^2 = \left(\frac{1}{R^2} - \frac{1}{R'^2}\right)\left(\frac{1}{r^2} - \frac{1}{s^2}\right).$$

182. Let $\phi\,(x,\,y,\,z) = 0$ be the equation to a surface of the second order; if tangent lines be drawn to it from the point $(a,\,\beta,\,\gamma)$, shew that the equation to the plane which contains all the points of contact is

$$2u + (x - a)\frac{du}{da} + (y - \beta)\frac{du}{d\beta} + (z - \gamma)\frac{du}{d\gamma} = 0,$$

where $u = \phi\,(a,\,\beta,\,\gamma)$.

183. Let $\phi(x, y, z) = 0$ be the equation to a surface of the second order; then the equation to the enveloping cone which has its vertex at the point (α, β, γ) is

$$2u\left\{(x-\alpha)^2\frac{d^2u}{d\alpha^2} + (y-\beta)^2\frac{d^2u}{d\beta^2} + (z-\gamma)^2\frac{d^2u}{d\gamma^2}\right.$$

$$\left. + 2(x-\alpha)(y-\beta)\frac{d^2u}{d\alpha d\beta} + 2(y-\beta)(z-\gamma)\frac{d^2u}{d\beta d\gamma} + 2(z-\gamma)(x-\alpha)\frac{d^2u}{d\gamma d\alpha}\right\}$$

$$= \left\{(x-\alpha)\frac{du}{d\alpha} + (y-\beta)\frac{du}{d\beta} + (z-\gamma)\frac{du}{d\gamma}\right\}^2,$$

where $u = \phi(\alpha, \beta, \gamma)$.

Example. Determine the enveloping cone of the surface

$$\frac{x^2}{a^2} + \frac{y^2}{b^2} + \frac{z^2}{c^2} = 1,$$

from the point (a, b, c) as vertex.

184. Let $\phi(x, y, z) = 0$ be the equation to a surface of the second order; then the equation to the enveloping cylinder which has its generating lines parallel to the straight line

$$\frac{x}{l} = \frac{y}{m} = \frac{z}{n} \text{ is}$$

$$2u\left\{l^2\frac{d^2u}{dx^2} + m^2\frac{d^2u}{dy^2} + n^2\frac{d^2u}{dz^2} + 2lm\frac{d^2u}{dxdy} + 2mn\frac{d^2u}{dydz} + 2nl\frac{d^2u}{dzdx}\right\}$$

$$= \left(l\frac{du}{dx} + m\frac{du}{dy} + n\frac{du}{dz}\right)^2,$$

where $u = \phi(x, y, z)$.

Example. Determine the enveloping cylinder of the surface $\frac{x^2}{a^2} + \frac{y^2}{b^2} + \frac{z^2}{c^2} = 1$, which has its generating lines parallel to the straight line $\frac{x}{a} = \frac{y}{b} = \frac{z}{c}$.

185. Shew that the surface

$$(ax + by + cz - 1)^2 + 2a' yz + 2b' zx + 2c' xy = 0$$

touches the co-ordinate axes; and find the equation to the cone which has its vertex at the origin and passes through the curve of section of the surface by the plane through the points of contact.

186. Straight lines are drawn through the origin perpendicular to the tangent planes to the cone

$$ax^2 + by^2 + cz^2 + 2a'yz + 2b'zx + 2c'xy = 0:$$

shew that they will generate the cone which has for its equation

$$(bc - a'^2)x^2 + (ca - b'^2)y^2 + (ab - c'^2)z^2$$
$$+ 2(b'c' - aa')yz + 2(c'a' - bb')zx + 2(a'b' - cc')xy = 0.$$

Shew also that if straight lines be drawn through the origin perpendicular to the tangent planes to the second cone they will generate the first cone.

III. *Surfaces in general.*

187. If a sphere pass through the origin of co-ordinates and its centre is on the surface defined by the equation

$$\{3\,(x^2 + y^2 + z^2) - b^2\}^2 = (x + y + z)^2\,(x^2 + y^2 + z^2),$$

the sum of the spherical surfaces cut off by the co-ordinate planes is constant and $= 2\pi b^2$.

188. A straight line drawn from the centre of an ellipsoid meets the ellipsoid at P and the sphere on the diameter $2a$ at Q : shew that the tangent planes at P and Q contain a constant angle α if the co-ordinates of P satisfy the equation

$$(x^2 + y^2 + z^2)\left(\frac{x^2}{a^4} + \frac{y^2}{b^4} + \frac{z^2}{c^4}\right) = \sec^2\alpha.$$

189. If a straight line be drawn from the centre of an hyperboloid whose equation is $\dfrac{x^2}{a^2} + \dfrac{y^2}{b^2} - \dfrac{z^2}{c^2} = 1$ to meet the surface at P, P', and a point Q be taken in CP produced such that $CP^2 = l\,(QP + QP')$, where l is a constant, shew that the locus of Q is defined by the equation

$$\left(\frac{x^2}{a^2} + \frac{y^2}{b^2} - \frac{z^2}{c^2}\right)^2 = \frac{x^2 + y^2 + z^2}{4l^4}.$$

190. The shadow of a given ellipsoid thrown by a luminous point on the plane which passes through two of the principal axes has its centre on the curve in which the same plane intersects the ellipsoid: shew that the equation to the locus of the luminous point is

$$\frac{x^2}{a^2} + \frac{y^2}{b^2} = \left(\frac{z^2}{c^2} - 1\right)^2.$$

191. If light fall from a luminous point whose co-ordinates are α, β, γ, on a surface whose equation is $xyz = m^3$, the boundary of light and shade lies on an hyperboloid of one or two sheets according as the product of α, β, γ, is negative or positive.

192. Tangent planes to an ellipsoid are drawn at a given distance from the centre: find the projections on the principal planes of the curve which is the locus of the points of contact. In what case will one of these projections consist of two straight lines?

193. Shew that the surfaces whose equations are

$$(x^2 + y^2)(a^2 - z^2) - c^2 y^2 = 0,$$

and

$$(x - c)^2 + y^2 + z^2 - a^2 = 0,$$

touch one another; and that the projection of the curve of contact on one of the co-ordinate planes is a circle, and on another a parabola.

194. Two circles have a common diameter AB and their planes are inclined to each other at a given angle; on PP', any chord of one of them parallel to AB, is described a circle with its plane parallel to that of the other circle: shew that the surface generated by these circles is an ellipsoid the squares of whose axes are in arithmetical progression.

195. Find the equation to the surface generated by straight lines drawn through the origin parallel to normals to the surface $\dfrac{x^2}{a^2} + \dfrac{y^2}{b^2} + \dfrac{z^2}{c^2} = 1$ at points where it is intersected by the surface $\dfrac{x^2}{a^2 - k^2} + \dfrac{y^2}{b^2 - k^2} + \dfrac{z^2}{c^2 - k^2} = 1$; k being greater than c and less than a.

196. An ellipse of given excentricity moving with its plane parallel to the plane of (y, z) and touching at the extremities of its axes the planes of (x, y) and (x, z) always passes through the curve whose equations are $y = x$, $cz = x^2$: find the equation to the surface generated, and determine the volume bounded by the surface and two given positions of the generating ellipse.

197. Shew that the surface determined by

$$z = F\left(\frac{ax + by + cz}{a'x + b'y + c'z}\right)$$

is cut by planes parallel to the plane of (x, y) in straight lines.

198.   A tangent line to the surface of the ellipsoid

$$\frac{x^2}{a^2} + \frac{y^2}{b^2} + \frac{z^2}{b^2} = 1$$

passes through the axis of z, and a given curve in the plane of (x, y): shew how to find the equation to the surface generated by it.   Example.   The curve being $\frac{x^2}{a^2} + \frac{y^2}{b^2} = m^2$, shew that the surface consists of two cones of the second order.

199.   Find the differential equation to the surface generated by a straight line which passes through two given curves and remains parallel to the plane of (x, y).   Shew that the equation $x = (y - b) \tan \frac{z}{a}$ represents such a surface.

200.   Determine the surface generated by a straight line which moves parallel to the plane of (x, y), and passes through the axis of z and through the curve given by

$$\frac{x^2}{a^2} + \frac{y^2}{b^2} + \frac{z^2}{c^2} = 1, \quad \frac{x}{a} + \frac{y}{b} = 1.$$

201.   Determine the surface generated by a straight line which moves parallel to the plane of (x, y) and passes through the following curves (1) and (2):

$$\frac{x^2}{a^2} + \frac{z^2}{c^2} = 1, \quad y = 0 \dots\dots\dots\dots\dots(1),$$

$$\frac{y^2}{b^2} + \frac{z^2}{c^2} = 1, \quad x = 0 \dots\dots\dots\dots\dots(2).$$

202.   A surface is generated by a straight line which always passes through the two fixed straight lines

$$y = mx, \ z = c; \text{ and } y = -mx, \ z = -c:$$

shew that the equation to the surface generated is of the form

$$\frac{mcx - yz}{c^2 - z^2} = f\left(\frac{mzx - cy}{c^2 - z^2}\right).$$

203. If a surface be such that at any point of it a straight line can be drawn lying wholly on the surface and intersecting the axis of z, then at every point of the surface

$$x^2 \frac{d^2z}{dx^2} + 2xy \frac{d^2z}{dxdy} + y^2 \frac{d^2z}{dy^2} = 0.$$

204. The general equation to surfaces generated by a straight line which is always parallel to the plane

$$lx + my + nz = 0,$$

is $\left(m + n\frac{dz}{dy}\right)^2 \frac{d^2z}{dx^2} - 2\left(m + n\frac{dz}{dy}\right)\left(l + n\frac{dz}{dx}\right)\frac{d^2z}{dxdy}$

$$+ \left(l + n\frac{dz}{dx}\right)^2 \frac{d^2z}{dy^2} = 0.$$

205. Find the equation to the surface generated by a straight line which always passes through each of two given straight lines in space, and also through the circumference of a circle whose plane is parallel to them both, and whose centre bisects the shortest distance between them.

206. A surface is generated by a straight line which always intersects the straight line $\dfrac{x-a}{l} = \dfrac{y-b}{m} = \dfrac{z-c}{n}$, and is parallel to the plane $\lambda x + \mu y + \nu z = 0$: find the functional equation to the surface; also find the differential equation.

207. Find the general functional equation to surfaces generated by the motion of a straight line which always intersects and is perpendicular to a given straight line.

If a surface whose equation referred to rectangular axes is

$$ax^2 + by^2 + cz^2 + 2a'yz + 2b'zx + 2c'xy + 2a''x + 2b''y + 2c''z + 1 = 0$$

be capable of generation in this manner, shew that

$$a + b + c = 0, \quad aa'^2 + bb'^2 + cc'^2 = 2a'b'c' + abc.$$

208. Shew that $xyz = c(x^2 - y^2)$ represents a concidal surface.

209. Shew that the condition in order that
$$ax^2 + by^2 + cz^2 + 2a''x + 2b''y + 2c''z = f$$
may represent a conical surface is
$$\frac{a''^2}{a} + \frac{b''^2}{b} + \frac{c''^2}{c} + f = 0.$$

210. Shew that in order that the equation
$$ax^2 + by^2 + cz^2 + 2a'yz + 2b'zx + 2c'xy + 2a''x + 2b''y + 2c''z + f = 0$$
may represent a cylindrical surface, it is necessary that
$$\frac{a''}{aa' - b'c'} + \frac{b''}{bb' - c'a'} + \frac{c''}{cc' - a'b'} = 0.$$
Is this condition sufficient as well as necessary ?

211. Explain the two methods of generating a developable surface; find the differential equation to developable surfaces from each mode of generation.

212. Are the following surfaces developable ?

(1) $xyz = a^3$;   (2) $z - c = \sqrt{(xy)}$.

213. Find the differential equation to a surface whose tangent plane at any point includes with the three co-ordinate planes a pyramid of constant volume; shew that the surface is generally developable, the only exception being the surface determined by $xyz = $ a constant.

214. Find the equation to the surface in which the tangent plane at (x, y, z) meets the axis of z at a distance from the origin equal to that of (x, y, z) from the origin.

215. Find the equation to the surface in which the tangent plane at (x, y, z) meets the axis of z at a distance $\dfrac{k^{n+1}}{z^n}$ from the origin. If $n = 1$, give that form to the arbitrary function which will produce the equation to an ellipsoid.

216. A plane passes through $(0, 0, c)$ and touches the circle
$$x^2 + y^2 = a^2, \quad z = 0 ;$$
determine the locus of the ultimate intersections of the plane.

217. Three points move with given uniform velocities along three rectangular axes from given positions: shew how to find the surface to which the plane passing through their contemporaneous positions is always a tangent.

218. Spheres of constant radius r are described passing through the origin: find the envelop of the planes of contact of tangent cones having a fixed vertex at the point (a, b, c).

219. Find the locus of the ultimate intersections of a series of planes touching two parabolas which lie in planes perpendicular to each other and have a common vertex and axis.

220. The sphere $x^2 + y^2 + z^2 = 2ax + 2by + 2cz$ is cut by another sphere which passes through the origin and has its centre on the surface

$$\frac{x^2}{a^2} + \frac{y^2}{b^2} + \frac{z^2}{c^2} = 1:$$

shew that the equation to the envelop of the planes of intersection is

$$\frac{1}{ax} + \frac{1}{by} + \frac{1}{cz} = 0.$$

221. From each point of the exterior of two concentric ellipsoids, whose axes are in the same directions, tangent planes are drawn to the surface of the interior ellipsoid: shew that all the planes of contact corresponding to the several points of the exterior surface touch another concentric ellipsoid.

222. From any point P in the surface of an ellipsoid straight lines are drawn so as each to pass through one of three conjugate diameters, and be parallel to the plane containing the other two; these straight lines meet the surface again at P_1, P_2, P_3: find the equation to the plane which passes through these points, and the locus of the ultimate intersections of all such planes, the diameters remaining fixed while P moves; and shew that its volume $= \dfrac{3}{35}$ of that of the ellipsoid.

223. A cone envelopes the ellipsoid

$$\frac{x^2}{a^2} + \frac{y^2}{b^2} + \frac{z^2}{c^2} = 1,$$

and its vertex moves on the similar ellipsoid

$$\frac{(x-\alpha)^2}{a^2} + \frac{(y-\beta)^2}{b^2} + \frac{(z-\gamma)^2}{c^2} = n^2:$$

shew that the plane of contact touches the surface

$$\left(\frac{\alpha x}{a^2} + \frac{\beta y}{b^2} + \frac{\gamma z}{c^2} - 1\right)^2 = n^2\left(\frac{x^2}{a^2} + \frac{y^2}{b^2} + \frac{z^2}{c^2}\right).$$

224. A sphere is described having its centre at a point on the surface of an elliptic paraboloid, and its radius equal to the distance of the point from a fixed plane perpendicular to the axis of the paraboloid: determine the envelop of all such spheres.

225. One axis of an ellipsoid is gradually increased while the volume remains constant, and the section containing the other axes similar: find the enveloping surface.

226. Shew that an infinite number of plane centric sections of an hyperboloid of one sheet may be drawn, each possessing the following property, namely that the normals to the surface at the curve of section all pass through one or other of two straight lines lying in the same plane with the two possible axes.

Shew that these centric planes envelope the asymptotic cone.

227. Shew that the envelop of a sphere of which any one of one series of circular sections of an ellipsoid is a diametral plane, is another ellipsoid touching a sphere described on the mean axis of the former ellipsoid as diameter, in a plane perpendicular to the straight line which passes through the centres of the series of circular sections.

228. Find the envelop of a sphere of constant radius having its centre on a given circle, and determine the section

by a tangent plane to the envelop perpendicular to the plane of the circle.

229. From every point of the surface

$$4z\,(lx^2 + l'y^2) + ll'\,(x^2 + y^2) = 0\dots\dots\dots\dots\dots(1),$$

as vertex, enveloping cones are drawn to the paraboloid

$$ll'z - (lx^2 + l'y^2) = 0\dots\dots\dots\dots\dots\dots(2).$$

Shew that the planes of contact all touch the surface

$$4z\,(lx^2 + l'y^2) - (l^2x^2 + l'^2y^2) = 0.$$

Also shew that the surface (1) contains the directrices of all the sections of (2) made by planes through its axis.

IV. *Curves.*

230. A curve is determined by the equations $x + z = a$, and $x^2 + y^2 + z^2 = a^2$: find the points where the tangent makes an angle β with the axis of z.

231. Find the equations to the tangent to the curve

$$\frac{x^2}{a^2} + \frac{y^2}{b^2} - \frac{z^2}{c^2} = 0, \quad \frac{x^2}{a^2} + \frac{y^2}{b^2} + \frac{z^3}{c^2} = 1.$$

232. Find the equations to the tangent to the curve

$$y^2 = ax - x^2, \quad z^2 = a^2 - ax.$$

233. Find the equations to the curve traced out on the surface

$$\frac{x^2}{a} + \frac{y^2}{b} = z$$

by the extremities of the latera recta of sections made by planes passing through the axis of z.

234. Having given the equations to a curve in space referred to three rectangular axes, find the length of the perpendicular from the origin upon the tangent at any point.

Example. $x = a \cos \theta, \quad y = a \sin \theta, \quad z = \frac{a}{2}(e^{\theta} + e^{-\theta})$.

Shew that if the perpendicular be invariable either the curve or its involute lies on the surface of a sphere.

235. A perpendicular is drawn from the origin on the tangent to a curve at the point (x, y, z). If x', y', z' are the co-ordinates of the foot of the perpendicular, shew that

$$x' = x - \frac{dx}{ds}\left(x\frac{dx}{ds} + y\frac{dy}{ds} + z\frac{dz}{ds}\right),$$

$$y' = y - \frac{dy}{ds}\left(x\frac{dx}{ds} + y\frac{dy}{ds} + z\frac{dz}{ds}\right),$$

$$z' = z - \frac{dz}{ds}\left(x\frac{dx}{ds} + y\frac{dy}{ds} + z\frac{dz}{ds}\right).$$

236. Shew how to determine the locus of the feet of the perpendiculars from the origin upon the tangent lines to a given curve.

The equations to the curve being $x = a \cos \theta$, $y = a \sin \theta$, $z = c\theta$, shew that the locus is a curve lying on the surface

$$\frac{x^2 + y^2}{c^2} - \frac{z^2}{a^2} = \frac{a^2}{c^2}.$$

237. A curve is traced on a sphere so that its tangent makes always a constant angle with a fixed plane: find its length from cusp to cusp.

238. Find the equation to the normal plane to a curve at any proposed point; if the normal plane always passes through a fixed straight line, shew that the curve is a circle.

239. Find the direction cosines of the straight line in which the normal plane at any point of a curve meets the osculating plane at that point.

240. Find the equation to the osculating plane at any point of the curve

$$x = a \cos t, \quad y = b \sin t, \quad z = ct.$$

241. If (x, y, z) be any point on a curve, shew what plane is represented by the equation

$$(x' - x) \frac{d^2x}{ds^2} + (y' - y) \frac{d^2y}{ds^2} + (z' - z) \frac{d^2z}{ds^2} = 1.$$

242. Find the value of the radius of absolute curvature from the consideration that it is the perpendicular from (x, y, z) on the plane in the preceding Example.

243. Find the radius of curvature of the curve

$$\frac{x}{h} + \frac{y}{k} = 1, \quad x^2 + z^2 = a^2,$$

at any point.

244. Determine the radii of absolute and spherical curvature and the co-ordinates of the centres in the case of a helix.

245. Find the radius of absolute curvature at a given point in the curve determined by

$$x = a \cos \theta, \quad y = b \sin \theta, \quad z = c\theta.$$

246. Required the locus from which three given spheres will appear of equal magnitudes.

247. A triangular pyramid upon a given base is such that given lengths being measured along the three edges from the base the remainders of the edges are always equal to one another: shew that the locus of the vertex is a conic section.

248. A point moves on the cone $y^2 + z^2 = m^2 x^2$ so that the tangent to its path is inclined to the axis of the cone at a constant angle β: shew that the locus of the point is determined by the equation to the cone together with the equation

$$k \log \frac{x}{c} = \sin^{-1} \frac{y}{mx},$$

where $k = \dfrac{\sqrt{(\tan^2 \beta - m^2)}}{m}$, and c is a constant.

249. Shew that a curve drawn on the surface of a right cone so as to cut the generating lines at a constant angle is a curve of constant inclination to the base of the cone, and that the consecutive distances between the points at which it cuts any one generator are in geometrical progression.

250. From the vertex of a right cone two curves are drawn on the surface cutting all the generating lines at constant angles which are complementary: shew that the sum of the inverse squares of the arcs intercepted between the vertex and a given circular section is independent of the magnitudes of the complementary angles.

251. Find the curves of greatest inclination to the co-ordinate planes on the surface of which the equation is $xyz = a^3$.

252. Find the curves of greatest inclination to the plane of (x, y) on the surface $cz = xy$.

253. Tangent planes are drawn to the surface $cz = xy$ at all points where this surface is intersected by the cylinder $x^2 = ay$: find the equations to the edge of regression of the developable surface formed by the intersection of these tangent planes.

254. Find the angle between the osculating planes at two consecutive points of a curve.

255. The shortest distance between the tangents at two consecutive points of a curve of double curvature is $\dfrac{\delta\phi\,(\delta s)^2}{12\rho}$, where δs is the length of the arc between the two points, $\delta\phi$ is the angle between the osculating planes at the two points, and ρ is the radius of absolute curvature.

256. Shew that the edge of the developable surface formed by the locus of the ultimate intersections of normal planes of a curve of double curvature is the locus of the centres of spherical curvature of the curve.

Find the locus of the centres of spherical curvature of a helix.

257. Find the equation to the surface on which lie all the evolutes of the curve $x^2 = az$, $x = y$.

258. Shew that the curve represented by the equations

$$x^2 + y^2 + z^2 = a^2,$$

$$\sqrt{x} + \sqrt{y} + \sqrt{z} = c\,;$$

is cut perpendicularly by each of the series of surfaces

$$x^{\frac{3}{2}} + (\mu - 1)\,y^{\frac{3}{2}} = \mu z^{\frac{3}{2}},$$

μ being an arbitrary parameter.

259. A curve is traced on a surface: shew that the radius of absolute curvature at any point of this curve is the same as the radius of curvature of the section of the surface made by the osculating plane of the curve at that point.

260. A curve is traced on a sphere: shew that generally the radius of the sphere is the radius of spherical curvature of the curve; but that this does not hold if the curve be a *plane* curve, or plane for an indefinitely short period at any point.

261. If the normal plane of a curve constantly touches a given sphere the curve is rectifiable.

262. If (x', y', z') be the point in the locus of the centres of curvature corresponding to the point (x, y, z) on a curve, ρ the radius of curvature at (x, y, z), and s' the arc of the above locus, shew that $\dfrac{d\rho}{ds'}$ is the cosine of the angle between the tangent at (x', y', z') and the direction of ρ.

263. Find the equation to the surface on which lie all the evolutes to the curve formed by the intersection of the surfaces $y^2 = 4a(x+z)$, $z^2 = 4a(x+y)$; and determine the equations to that evolute which cuts the axis of x at a distance $7a$ from the origin.

V. *Curvature of Surfaces.*

264. Determine the conditions necessary in order that the surfaces whose equations are

$$ax^2 + by^2 + cz^2 + 2a'yz + 2b'zx + 2c'xy + 2z = 0,$$
$$Ax^2 + By^2 + Cz^2 + 2A'yz + 2B'zx + 2C'xy + 2z = 0,$$

may have their principal radii of curvature at the origin equal; and shew that if these conditions be fulfilled any sections of the two surfaces parallel to the plane of (x, y) will be similar.

265. Obtain the quadratic equation for determining the principal radii of curvature at any point of the surface

$$\phi(x) + \chi(y) + \psi(z) = 0 ;$$

and find the condition that the principal curvatures may be equal and opposite.

266. The locus of the points on the hyperboloid

$$\frac{x^2}{a^2} + \frac{y^2}{b^2} - \frac{z^2}{c^2} = 1,$$

for which the principal curvatures are equal and opposite, has for its projection on the plane of (x, y) the ellipse

$$\frac{x^2}{a^2}(a^2 + c^2) + \frac{y^2}{b^2}(b^2 + c^2) = a^2 + b^2.$$

267. Shew that the principal radii of curvature are equal in magnitude and opposite in sign at every point of the surface determined by

$$z = (y - b) \tan \frac{x}{a}.$$

268. The trace on the plane of (y, z) of the locus of the extremities of the principal radii of curvature of the ellipsoid whose equation is

$$\frac{x^2}{a^2} + \frac{y^2}{b^2} + \frac{z^2}{c^2} = 1$$

is given by the equation

$$\{(by)^{\frac{2}{3}} + (cz)^{\frac{2}{3}} - (b^2 - c^2)^{\frac{2}{3}}\} \left\{ \frac{b^2 y^2}{(a^2 - b^2)^2} + \frac{c^2 z^2}{(a^2 - c^2)^2} - 1 \right\} = 0.$$

260. Among the surfaces included in the equation

$$x\frac{dz}{dx} + y\frac{dz}{dy} = 0,$$

find that in which the principal radii of curvature are equal but of opposite sign.

270. Find the surface of revolution at every point of which the radii of curvature are equal in magnitude and opposite in sign.

271. If a, b be the principal radii of curvature at any point of a surface referred to the tangent plane at that point as the plane of (x, y) and the principal planes as planes of (x, z) and (y, z), then will the locus of the circles of curvature of all normal sections of the surface at the origin be

$$(x^2 + y^2 + z^2)\left(\frac{x^2}{a} + \frac{y^2}{b}\right) = 2z(x^2 + y^2).$$

272. Find the radius of curvature in any normal section of the surface

$$Ax^2 + By^2 + Cz^2 + 2A'yz + 2B'zx + 2C'xy + Ez = 0,$$

at the origin; and shew that the sum of the reciprocals of the radii of curvature in sections at right angles to each other is constant.

273. Required the sum of the principal radii of curvature at any point of a curved surface in terms of the co-ordinates of that point.

The equation to the surface being $f(x, y, z) = 0$, express the result by partial differential coefficients of $f(x, y, z)$.

274. If $r = f(\theta, \phi)$ be the equation to a surface referred to the tangent plane at the origin as the plane of (x, y), then

the radius of curvature at the origin of a normal section inclined at any angle ϕ to the plane of (x, z) is

$$= -\frac{1}{2} \text{ limit of } \frac{dr}{d\theta} \text{ when } \theta = \frac{\pi}{2}.$$

Example. In the surface $xy = az$, shew that $\rho = \dfrac{a}{\sin 2\phi}$.

275. If the surface $(x - a)^2 + (y - b)^2 = (z - c)^2$ have contact of the second order with the surface $z = f(x, y)$, shew that at the point of contact

$$\frac{p^2 - 1}{r} = \frac{q^2 - 1}{t} = \frac{pq}{s}.$$

276. At either point at which the surface

$$\left(\frac{x^2}{a^2} - 1\right)^2 + \left(\frac{y^2}{b^2} - 1\right)^2 = \frac{z^2}{c^2} + 1$$

meets the axis of z, an elliptic paraboloid may be found, which has at its vertex a complete contact of the third order with the surface.

277. Find the radius of curvature of a normal section of an ellipsoid of revolution made by a plane inclined to the meridian at any given angle.

278. Shew that the locus of the focus of an ellipse rolling along a straight line is a curve such that if it revolves about that straight line, the sum of the curvatures of any two normal sections at right angles to one another will be the same for all points of the surface generated.

279. In any surface of the second order the tangents to the lines of curvature at any point are parallel to the axes of any plane section parallel to the tangent plane to the surface at that point.

280. Obtain the differential equation to the projection on one of the co-ordinate planes of the lines of curvature of a surface. Apply the equation to determine the lines of curvature of a surface of revolution.

281. Determine the lines of curvature on the surface

$$xy = az.$$

282. Find the radius of curvature of any normal section of a surface at a given point.

If $y' - y = m(x' - x)$ be the projection on the plane of (x, y) of the tangent line to the curve of section, shew that the values of m corresponding to the principal sections of the surface $\dfrac{y}{z} = f\left(\dfrac{x}{z}\right)$ at the point (x, y, z) are

$$\frac{y}{x} \text{ and } -\frac{x + pz}{y + qz}.$$

283. The links of a chain are circular, being of the form of the surface generated by the revolution of a circle whose radius is one inch about a line in its own plane at a distance of four inches from the centre : apply Euler's formula for the curvature of surfaces to shew that if one link be fixed, the next cannot be twisted through an angle greater than 60° without shortening the chain.

284. An annular surface is generated by the revolution of a circle about an axis in its own plane: shew that one of the principal radii of curvature at any point of the surface varies as the ratio of the distance of this point from the axis to its distance from the cylindrical surface described about the axis and passing through the centre of the circle.

285. If ρ, ρ' be the greatest and least radii of curvature of a curve surface at a given point, ϕ, ψ the angles which the normal to the surface at a given point makes with the axes of x and y, shew that

$$\frac{1}{\rho} + \frac{1}{\rho'} = \frac{d}{dx}\cos\phi + \frac{d}{dy}\cos\psi.$$

286. Define an umbilicus. In what sense do you understand that there is an infinite number of lines of curvature at an umbilicus? And from this consideration deduce the partial differential equations which exist at such points.

287. Shew that in the surface

$$\phi(x) + \chi(y) + \psi(z) = 0,$$

if $\phi''(x) = \chi''(y) = \psi''(z)$ at any point, that point is an umbilicus.

288. Find the umbilici of the surface

$$2z = \frac{x^2}{\alpha} + \frac{y^2}{\beta}.$$

289. Find the umbilici of the surface $xyz = a^3$.

290. Shew that the radius of normal curvature of the surface $xyz = a^3$ at an umbilicus is equal to the distance of the umbilicus from the origin of co-ordinates.

291. A sphere described from the origin with radius $\dfrac{abc}{ac + ab + bc}$ will touch the surface

$$\left(\frac{x}{a}\right)^{\frac{2}{3}} + \left(\frac{y}{b}\right)^{\frac{2}{3}} + \left(\frac{z}{c}\right)^{\frac{2}{3}} = 1$$

in points which are umbilici.

292. Shew that a sphere whose centre is at the origin and whose radius r is determined by the equation

$$r^{\frac{2n}{n-2}} = a^{\frac{2n}{n-2}} + b^{\frac{2n}{n-2}} + c^{\frac{2n}{n-2}}$$

will touch the surface

$$\left(\frac{x}{a}\right)^{n} + \left(\frac{y}{b}\right)^{n} + \left(\frac{z}{c}\right)^{n} = 1$$

at umbilici; and the radius of normal curvature of the surface at these points is $\dfrac{r}{n-1}$.

293. Determine whether there is an umbilicus on the surface

$$(x^2 + y^2 + z^2)^2 = 4a^2(x^2 + y^2).$$

294. If R be the radius of absolute curvature at any point of a curve defined by the intersection of two surfaces $u_1 = 0$, $u_2 = 0$, and r_1, r_2 be the radii of curvature of the sections of $u_1 = 0$, $u_2 = 0$, made by the tangent planes to $u_2 = 0$, $u_1 = 0$, respectively at that point, shew that R, r_1, r_2 will be connected by the relation

$$\frac{1}{R^2} = \frac{1}{r_1^{\,2}} - \frac{2\cos\theta}{r_1 r_2} + \frac{1}{r_2^{\,2}},$$

θ being the angle between the tangent planes.

295. Two surfaces touch each other at the point P; if the principal curvatures of the first surface at P be denoted by $a \pm b$, those of the second by $a' \pm b'$, and if ϖ be the angle between the principal planes to which $a + b$, $a' + b'$, refer, and δ the angle between the two branches at P of the curve of intersection of the surfaces, shew that

$$\cos^2 \delta = \frac{(a - a')^2}{b^2 + b'^2 - 2bb'\cos 2\varpi}.$$

296. Shew that at any point of a developable surface, the curvature of any normal section varies as the square of the sine of the angle which this section makes with the generating line, and that at different points along the same generating line the principal radius of curvature varies as the distance from the point of intersection of consecutive generating lines.

297. A surface is generated by the motion of a straight line which always intersects a fixed axis. If P be any point in this axis at a distance x from the origin, ϕ the angle which the generating line through this point makes with the axis, and ψ the angle which the plane through the axis and the generating line makes with its initial position, shew that the principal radii of curvature at P are

$$\cot\frac{\phi}{2}\frac{dx}{d\psi} \text{ and } -\tan\frac{\phi}{2}\frac{dx}{d\psi}.$$

298. The shortest distance between two points on a curved surface never coincides with a line of curvature unless it be a plane curve.

VI. *Miscellaneous Examples.*

299. Find the condition which must hold in order that the equations

$$ax + c'y + b'z = 0, \quad by + a'z + c'x = 0, \quad cz + b'x + a'y = 0,$$

may represent a straight line; and shew that in that case the straight line is determined by the following equations:

$$x\,(aa' - b'c') = y\,(bb' - c'a') = z\,(cc' - a'b').$$

300. Shew that the six planes which bisect the interior angles of a tetrahedron meet at a point.

301. Shew that the three planes which bisect the exterior angles round one face of a tetrahedron and the three planes which bisect the interior angles formed by the other three faces meet at a point.

302. If $\alpha = 0$, $\beta = 0$, $\gamma = 0$, $\delta = 0$ be the equations to the faces of a tetrahedron expressed in a suitable form, A, B, C, D the areas of the respective faces, shew that

$$A\alpha + B\beta + C\gamma + D\delta = \text{a constant.}$$

Interpret $\qquad A\alpha + B\beta + C\gamma = 0.$

303. If $\alpha = 0$, $\beta = 0$, $\gamma = 0$ be the equations to three planes which form a trihedral angle, the equation to a cone of the second degree which has its vertex at the angular point and touches two of the planes at their intersections with the third, is $\gamma^2 - k\alpha\beta = 0$.

304. Give the geometrical interpretation of the equation $uu' = kvv'$, where k is a constant, and the other letters denote linear functions of x, y, z. Hence shew that there must exist surfaces of the second order which contain straight lines.

305. If $u_1 = 0$, $u_2 = 0$, $u_3 = 0$ be the equations to three planes, interpret the equation

$$\sqrt{\left(\frac{u_1}{a}\right)} + \sqrt{\left(\frac{u_2}{b}\right)} + \sqrt{\left(\frac{u_3}{c}\right)} = 0.$$

T. A. G. $\qquad\qquad\qquad$ 4

306. Also interpret $Au_2u_3 + Bu_1u_3 + Cu_1u_2 = 0$.

307. If u, v, w are linear functions of x, y, z, shew that $uv = w^2$ represents a conical surface; and shew that the equation to the tangent plane is $\lambda^2 u - 2\lambda w + v = 0$.

308. Let $v = 0$ be the equation to a surface of the second order; $u_1 = 0$ and $u_2 = 0$ the equations to two planes: shew that by giving a suitable value to the constant λ the equation $v + \lambda u_1 u_2 = 0$ will represent *any* surface of the second order which passes through the intersections of the two planes with the given surface.

309. If $t = 0$, $u = 0$, $v = 0$, $w = 0$ be the equations to four given planes, and λ, μ be two arbitrary constants, shew that $t + \lambda u = 0$ represents a plane which passes through a fixed straight line, $t + \lambda u + \mu v = 0$ a plane which passes through a fixed point, and $tw + \mu uv = 0$ a surface of the second order which contains four fixed straight lines.

310. If the equation to a surface of the second order be $v_2 + 2u_1 + 1 = 0$, where u_1 and u_2 represent the terms of the first and second order respectively, and tangent planes be drawn to the surface from any point of the plane determined by $u_1 + 1 = 0$, the planes of contact will all pass through the origin.

311. If α, β, γ, δ be the distances of any point from the faces of a tetrahedron, shew that the general equation to a surface of the second order circumscribing the tetrahedron is

$$A\alpha\beta + B\alpha\gamma + C\alpha\delta + D\beta\gamma + E\beta\delta + F\gamma\delta = 0.$$

Determine the condition necessary in order that the straight line $\dfrac{\alpha}{l} = \dfrac{\beta}{m} = \dfrac{\gamma}{n}$ may touch the surface; and hence shew that the equation to the tangent plane at the point $(\alpha = \beta = \gamma = 0)$ is

$$C\alpha + E\beta + F\gamma = 0.$$

312. If A, B, C, D be the angular points of a tetrahedron; α, β, γ, δ the distances of any point from the faces respectively

opposite to them, shew that the general equation of the hyperboloid of one sheet of which AB, CD are generating lines is

$$l\alpha\gamma + m\alpha\delta + n\beta\gamma + p\beta\delta = 0.$$

313. Interpret $u^2 + v^2 + w^2 = a$, where u, v, w are linear functions of x, y, z, and a is a constant.

314. Interpret $u^2 + v^2 - w^2 = a$.

315. Interpret $u^2 + v^2 = a$.

316. Interpret $u^2 - v^2 = a$.

317. Interpret $u^2 = a$.

318. Interpret $u^2 + v^2 + w = 0$.

319. Interpret $u^2 - v^2 + w = 0$.

320. Interpret $u^2 + v = 0$.

321. Interpret $\theta = $ constant.

322. Interpret $\phi = $ constant.

323. Interpret $r = $ constant.

324. Interpret $F(\theta) = 0$.

325. Interpret $F(\phi) = 0$.

326. Interpret $F(r) = 0$.

327. Interpret $F(r, \theta) = 0$.

328. Interpret $F(r, \phi) = 0$.

329. Interpret $F(\theta, \phi) = 0$.

330. Describe the form of the surface represented by

$$\frac{\theta^2}{\alpha^2} + \frac{\phi^2}{\beta^2} = 1.$$

331. Interpret $r = a \cos \theta$.

332. Shew that the polar equation to a plane may be put in the form $\dfrac{c}{r} = \cos \alpha \cos \theta + \sin \alpha \sin \theta \cos (\phi - \beta)$.

4—2

333. If p be the perpendicular from the origin on the tangent plane at any point of the surface $r = f(\theta, \phi)$,

$$\frac{1}{p^2} = \frac{1}{r^2} + \frac{1}{r^4}\left(\frac{dr}{d\theta}\right)^2 + \frac{\cosec^2\theta}{r^4}\left(\frac{dr}{d\phi}\right)^2.$$

334. Shew that the polar partial differential equation to conical surfaces with the origin as vertex is

$$\frac{d\phi}{dr} = 0, \quad \text{or } \frac{d\theta}{dr} = 0.$$

335. Shew that the polar partial differential equation to surfaces of revolution is

$$\frac{dr}{d\phi} = 0.$$

336. Shew that the polar partial differential equation to cylindrical surfaces is

$$\sin\theta \frac{dr}{d\theta} + r\cos\theta = 0.$$

337. Find the volume of a pyramid which has its vertex at the point (x, y, z), and for its base the triangle formed by joining the points where the plane

$$\frac{x}{a} + \frac{y}{b} + \frac{z}{c} = 1$$

meets the co-ordinate axes.

338. Demonstrate the formulæ for transforming from one set of rectangular axes to another. Shew that all six axes lie on a certain cone of the second order.

339. If two systems of rectangular axes have the same origin, and (a_1, b_1, c_1), (a_2, b_2, c_2), (a_3, b_3, c_3) be the direction cosines of one system with respect to the other, shew that

$$a_1^2 = (b_2 c_3 - b_3 c_2)^2, \quad a_2^2 = (b_3 c_1 - b_1 c_3)^2, \quad a_3^2 = (b_1 c_2 - b_2 c_1)^2.$$

340. Find the size of a cube which will stop up a tube of uniform bore, the section of which is a regular hexagon whose sides are given.

341. The stereographic projection of a cube on a plane perpendicular to its diagonal (the pole being in the diagonal produced) is an equilateral hexagon, whose angles are alternately greater and less. If the pole of projection be at a distance from the cube equal to its diagonal, the sines of two adjacent angles of the hexagon are as 8 to 5.

342. A cube being placed with one face in contact with a given plane, determine the position of a luminous point such that the shadow cast on the plane shall be an equilateral pentagon of which the diagonal of the above face is one side.

343. Find the locus of the centres of the sections of a surface of the second order made by planes which all pass through a fixed straight line.

344. Find the locus of the middle points of all chords drawn in a surface of the second degree, the length of each chord varying as the diameter parallel to it.

345. From different points of the straight line $\dfrac{x}{a} = \dfrac{y}{b}$, $z = 0$, asymptotic straight lines are drawn to the hyperboloid

$$\frac{x^2}{a^2} + \frac{y^2}{b^2} - \frac{z^2}{c^2} = 1 \, ;$$

shew that they will all lie in the planes

$$\frac{x}{a} - \frac{y}{b} = \pm \frac{z}{c} \sqrt{2}.$$

346. Common tangent planes are drawn to the ellipsoids

$$\frac{x^2}{a^2} + \frac{y^2}{b^2} + \frac{z^2}{c^2} = 1, \text{ and } \frac{x^2}{a'^2} + \frac{y^2}{b'^2} + \frac{z^2}{c'^2} = 1 :$$

shew that the perpendiculars upon them from the origin lie in the surface of the cone

$$(a^2 - a'^2)\, x^2 + (b^2 - b'^2)\, y^2 + (c^2 - c'^2)\, z^2 = 0.$$

347. Determine what must be the form of a column in order that it may appear to be of uniform thickness to an observer in a given position.

348. If the section of the surface whose equation is

$$ax^2 + by^2 + cz^2 + 2a'yz + 2b'zx + 2c'xy + 2a''x + 2b''y + 2c''z + f = 0,$$

by the plane whose equation is

$$lx + my + nz = p,$$

be a rectangular hyperbola, then will

$$l^2(b+c) + m^2(c+a) + n^2(a+b) = 2a'mn + 2b'nl + 2c'lm.$$

349. If the section of the surface $xy + yz + zx = a^2$ by the plane $lx + my + nz = \delta$ be a parabola, then will

$$l^2 + m^2 + n^2 - 2mn - 2nl - 2lm = 0.$$

350. Shew that the asymptotes of any plane section of an hyperboloid are parallel to the straight lines in which the asymptotic cone is cut by the parallel central plane.

351. Two hyperboloids of one and two sheets respectively have the same asymptotic cone : shew that the sections of the hyperboloid of one sheet by tangent planes to the other hyperboloid will be ellipses of constant area if the points of contact lie on the curve of section of the surface by a certain concentric ellipsoid whose axes are in the ratio of the squares of the axes of the hyperboloids.

352. Through two straight lines given in space two planes are drawn so as to be at right angles : find the locus of their intersection.

353. A surface is generated by the intersection of planes cutting the axes of co-ordinates at distances from the origin equal to the respective co-ordinates of each point of the ellipsoid $\dfrac{x^2}{a^2} + \dfrac{y^2}{b^2} + \dfrac{z^2}{c^2} = 1$; if the surface so generated be treated as the ellipsoid and the process repeated n times, the equation to the nth surface will be

$$\left(\frac{x}{a}\right)^{\frac{2}{2n+1}} + \left(\frac{y}{b}\right)^{\frac{2}{2n+1}} + \left(\frac{z}{c}\right)^{\frac{2}{2n+1}} = 1.$$

354. A surface is generated by a variable circle whose plane is parallel to $x + y = 0$, and which always passes through the axes of x and y and the line $y = x$, $z = c$: find the equation to the surface. Also find the volume between the origin and the plane $x + y = c$.

355. Two equal parabolas have their common vertex at the origin, their axes coincident with the axis of x in opposite directions, and their planes coincident with the planes of (x, y) and (x, z) respectively; a straight line is drawn intersecting these curves and parallel to the plane $y = z$: find the locus of its trace on the plane of (y, z).

356. Explain the nature of the surface defined by the equation

$$\frac{y}{b} = \left\{ \left(\frac{x^2}{a^2} + \frac{z^2}{c^2} \right)^{\frac{1}{2}} - 1 \right\} \left\{ \left(\frac{x^2}{a^2} + \frac{z^2}{c^2} \right)^{\frac{1}{2}} - m \right\}^{\frac{1}{2}}$$

near the points where it meets the hyperboloid

$$\frac{x^2}{a^2} - \frac{y^2}{b^2} + \frac{z^2}{c^2} = 1$$

in the several cases in which m is $> =$ or < 1.

357. Planes are drawn perpendicular to the tangent lines to the surface $f(x, y, z) = 0$ at a point (x, y, z) in it: shew that if at that point

$$\frac{df}{dx} = 0, \qquad \frac{df}{dy} = 0, \qquad \frac{df}{dz} = 0,$$

$$\frac{d^2 f}{dx^2} = u, \; \dots \; \frac{d^2 f}{dy \, dz} = u', \dots$$

the locus of the ultimate intersections of the planes is the cone

$$(vw - u'^2)(\xi - x)^2 + \dots + 2(v'w' - uu')(\eta - y)(\zeta - z) + \dots = 0.$$

358. Shew that the normal cone at a singular point of the surface

$$(x^2 + y^2 + z^2)(a^2 x^2 + b^2 y^2 + c^2 z^2) - a^2(b^2 + c^2)x^2 - b^2(c^2 + a^2)y^2 - c^2(a^2 + b^2)z^2$$
$$+ a^2 b^2 c^2 = 0$$

has for its equation

$$(b^2 - c^2)(\xi - x_0)^2 + (c^2 - a^2)\eta^2 + (a^2 - b^2)(\zeta - z_0)^2$$
$$- \frac{c^2 + a^2}{ca} \sqrt{(a^2 - b^2)} \sqrt{(b^2 - c^2)}(\zeta - z_0)(\xi - x_0) = 0,$$

where $\quad x_0 = \dfrac{c\sqrt{(a^2 - b^2)}}{\sqrt{(a^2 - c^2)}}, \quad y_0 = 0, \quad z_0 = \dfrac{a\sqrt{(b^2 - c^2)}}{\sqrt{(a^2 - c^2)}}$

are the co-ordinates of the singular point.

359.   If $f(x, y, z, c) = 0$ represent a system of surfaces for which c is a variable parameter, shew how to find the locus of the points of contact of tangent planes to each one of the system, each tangent plane passing through a fixed point.

Example. $x^2 + y^2 = 2c(z - c)$, each of the tangent planes passing through the origin.

360.   If PN be the normal at the point P of any surface, a, b the principal radii of curvature at P, r the radius of curvature of the normal section made by a plane inclined at an angle θ to the principal section to which a refers, $PQ = s$ an indefinitely small arc in this section, shew that if D be the minimum distance of the normals at P and Q, c the distance from P of their point of nearest approach,

$$D^2 = \frac{\sin^2\theta\cos^2\theta(a - b)^2 s^2}{b^2\cos^2\theta + a^2\sin^2\theta}, \quad c = \frac{ab(b\cos^2\theta + a\sin^2\theta)}{b^2\cos^2\theta + a^2\sin^2\theta}.$$

Express also $\dfrac{D^2}{s^2}$ and c in terms of a, b and r.

361.   A family of surfaces is defined by the equation

$$F\{x, y, z, a, \phi(a), \psi(a)\} = 0,$$

where a is a variable parameter and $\phi(a)$ and $\psi(a)$ arbitrary functions of a: shew how to find the partial differential equation of the second order which belongs to the envelop, and prove that it assumes the form

$$rt - s^2 + Pr + Qs + Rt + S = 0,$$

where P, Q, R, S are functions of x, y, z, p and q.

362. If the equation to a system of surfaces contains an arbitrary parameter, shew that a curve can always be found which cuts them all at right angles.

363. Find the equations to a curve which cuts at right angles a series of ellipsoids which have their axes fixed in position and two of them of given length.

364. Shew that the section of a surface made by a plane parallel and indefinitely near to the tangent plane at any point in the immediate neighbourhood of that point is generally a conic section; and explain fully the peculiarity of the surface near a point where this conic section becomes two parallel straight lines.

365. Shew how to determine whether a curved surface has a tangent plane which touches it along a line. Examples:

(1) $(x^2 + y^2 + z^2)^2 = 4a^2 (x^2 + y^2)$.

(2) $b^3 z = (x^2 + y^2)^2 - a^2 (x^2 + y^2)$.

366. The extremities of the minor axis of the elliptic sections of a right cone made by parallel planes lie in two generating lines of the cone.

367. Shew how to cut from a given right cone an hyperbola whose asymptotes shall contain the greatest possible angle.

368. What is the section of a right cone by a plane when the cutting plane is parallel to a generating line, but not perpendicular to the plane containing the axis and that line?

369. Two spheres exterior to each other are inscribed in a right cone touching it in two circles on the same side of the vertex, and a plane is drawn touching the spheres and cutting the cone: shew that the section is an ellipse, that the points of contact of the spheres with the plane are the foci, and that the planes of the two circles contain the directrices.

370. A conical surface is placed with its circular base in contact with a plane. It is then slit along a generating line

and the vertex pressed in a direction perpendicular to the plane, the base remaining in contact while the surface opens out: shew that the extremities of the separating edges trace hyperbolic spirals.

371.   A triangle ABC revolves about a straight line which bisects the angle A, and the conical surface generated by the sides containing that angle is cut by a plane passing through the other side and perpendicular to the plane of the triangle in one of its positions: shew that if e be the excentricity of the section, $e^2 = \dfrac{(b-c)^2}{a^2}$.

372.   If the cone $x^2 + y^2 = z \, (mx + nz)$ cut any sphere which has its centre at the origin, shew that the projection on the plane of (x, z) of the curve of intersection is an hyperbola which has its centre at the origin.

373.   A straight inelastic band is wrapped smoothly on the surface of a cone : shew that however long it may be, the two ends of either of its edges cannot be made to meet, if the vertical angle of the cone be greater than 60°.

374.   Shew that an oblique cone on a circular base can be cut by a plane not parallel to the base, so that the section shall be a circle.   Shew that the cone so cut off is similar to the whole cone.

375.   If $f(x, y, z) = 0$ be the equation to any surface which passes through the origin, and $\phi(x, y, z)$ the sum of all the terms of lowest dimension in $f(x, y, z)$, shew that $\phi(x, y, z) = 0$ is the locus of all the tangent lines at the origin.

376.   Find the surface which touches at the origin the surface whose equation is

$$(x^2 + y^2 + z^2 + 2ax)^2 - 4(c^2 - a^2)(y^2 + z^2) = 0 \, ;$$

and shew that as a diminishes and ultimately vanishes, the tangent cone contracts and ultimately becomes a straight line, and that as a increases up to c it expands and finally becomes a plane.

377. A circle always touches the axis of z at the origin, and passes through a fixed straight line in the plane of (x, y): find the equation to the surface generated. Shew that the origin is a singular point, and that in its immediate neighbourhood the surface may be conceived to be generated by a circle having its plane parallel to that of (x, y), and its radius proportional to z^2.

378. A circle revolves round a chord in its own plane, the direction cosines of which are l, m, n: shew that the equation to the surface generated is

$$(lx + my + nz - h)^2 + [\sqrt{\{x^2 + y^2 + z^2 - (lx + my + nz)^2\}} - k]^2 = r^2,$$

where r is the radius of the circle; h and k the distances of the centre of the circle along and from the chordal line which is supposed to pass through the origin. Determine the singular points of this surface and the equation to the tangent cone there; also determine the equation to the tangent planes which touch the surface along circles.

379. The traces of a surface of the second order on two planes at right angles to each other are parabolas with their axes parallel to the line of intersection of the planes: find the condition that the surface may be a developable surface; and shew that in that case the trace of the surface on a plane perpendicular to the first two will also be a parabola.

380. Find the locus of the point at which a perpendicular from the origin meets a plane which cuts off a constant volume from the co-ordinate planes.

381. An oblique cylinder stands on a great circle of a sphere: determine the curve of intersection of the sphere and cylinder, and find the area of the spherical surface included within the cylinder.

382. The locus of the intersection of generating lines of an hyperboloid of one sheet which pass through extremities of conjugate diameters of the smallest elliptic section is a similar ellipse parallel to it, whose axes are to the axes of the former as $\sqrt{2}$ to 1.

383.   Find the surface which is generated by a straight line which is always parallel to the plane of (x, y), and passes through the axis of z, and also through the curve determined by $xyz = a^3$, $x^2 + y^2 = c^2$.

384.   A normal to the surface of which the equation is

$$x \cos nz - y \sin nz = 0,$$

moves along any one of its generating lines : determine the surface generated.

385.   The equations to a system of lines in space, straight or curved, contain two arbitrary parameters : shew how to find whether the lines can be cut at right angles by a system of surfaces, and when they can shew how to find the equation to that system.

Examples.   (1) Let the lines be a system of straight lines, each of which intersects two given straight lines which are perpendicular to each other but do not intersect.   (2) Let the equations to the system of lines be $A x^\alpha = B y^\beta = C z^\gamma$, where A, B, C, are arbitrary.

386.   The equation to a system of surfaces contains one arbitrary parameter ; normals are drawn at all points of one of these surfaces, and the lengths of the normals are taken proportional to the ultimate distances between the surface in question and a consecutive surface : shew how to find the equation to the locus of the extremities of the normals.

387.   The equations to a system of straight lines in space contain two arbitrary parameters : shew that when the roots of a certain quadratic are real and unequal, there are two planes passing through a given line of the system which contain consecutive lines.

388.   A cavity is just large enough to allow of the complete revolution of a circular disk of radius c, whose centre describes a circle of the same radius c, while the plane of the disk is constantly parallel to a fixed plane and perpendicular

to that of the circle in which its centre moves. Shew that the volume of the cavity is $\dfrac{2c^3}{3}(3\pi + 8)$.

389. The solid of which the surface is determined by the equation $\sqrt{x} + \sqrt{y} + \sqrt{z} = \sqrt{a}$ revolves round the fixed axis of z and makes for itself a cavity in a mass of yielding material: determine the form and magnitude of the cavity.

390. Shew that the surfaces represented by the equations

$$\frac{x^2}{a^2} + \frac{y^2}{b^2} + \frac{z^2}{c^2} = 1, \quad \frac{a^2}{x^2} + \frac{\beta^2}{y^2} + \frac{\gamma^2}{z^2} = 1,$$

will touch each other in eight points if

$$\frac{\alpha}{a} + \frac{\beta}{b} + \frac{\gamma}{c} = 1;$$

and prove that if tangent planes be drawn at these points they will form a solid whose opposite faces are similar and parallel, and the volume of which is

$$\frac{4}{3}\frac{(abc)^{\frac{3}{2}}}{(\alpha\beta\gamma)^{\frac{1}{2}}}.$$

391. Normals are drawn to the ellipsoid

$$\frac{x^2}{a^2} + \frac{y^2}{b^2} + \frac{z^2}{c^2} = 1$$

at every point of its curve of intersection with the sphere

$$x^2 + y^2 + z^2 = p^2:$$

shew that the equation to the curve in which the locus of these normals is cut by the plane of (y, z) is

$$\frac{b^2 y^2}{a^2 - b^2} + \frac{c^2 z^2}{a^2 - c^2} = a^2 - p^2.$$

392. Find the envelop of the plane

$$\alpha x + \beta y + \gamma z = 1,$$

the parameters α, β, γ being subject to the relations

$$a\alpha + b\beta + c\gamma = 1,$$

$$a^2\alpha^2 + b^2\beta^2 + c^2\gamma^2 = 1.$$

393. Determine the singular point on the surface

$$a^2x^2 + b^2y^2 = z^3(c - z),$$

and the locus of the tangent lines there.

394. The tangent plane to a surface S cuts an ellipsoid, and the locus of the vertex of the cone which touches the ellipsoid in the curve of intersection is another surface S'. Shew that S and S' are reciprocal; that is, that S may be generated from S' in the same manner as S' has been generated from S.

395. A plane moves so as always to cut off from an ellipsoid the same volume: shew that it will in every position touch a similar and concentric ellipsoid.

396. Shew that the tangent plane at any point of the surface

$$(ax)^2 + (by)^2 + (cz)^2 = 2(bcyz + cazx + abxy)$$

intersects the surface

$$ayz + bzx + cxy = 0$$

in two straight lines which are at right angles to one another.

397. A plane is drawn through the axis of y such that its trace upon the plane of (z, x) touches the two circles in which the plane of (z, x) meets the surface generated by the revolution round the axis of z of the circle

$$(x - a)^2 + z^2 = c^2,$$

where a is greater than c: determine the curve of intersection of the plane and the surface.

398. A straight line of constant length moves with one extremity in the axis, and the other in the surface of an elliptic cone: find the equation to the surface which is the locus of its middle point, and shew that the trace on a plane passing through the vertex at right angles to the axis is an ellipse.

399. A plane is drawn through a generating line of an hyperboloid of one sheet: shew that it meets the surface again in a straight line.

400. A plane moves so as always to enclose between itself and a given surface S a constant volume: shew that the envelop of the system of such planes is the same as the locus of the centres of gravity of the portions of the planes comprised within S.

401. OA, OB, OC are three straight lines mutually at right angles, and a luminous point is placed at C: shew that when the quantity of light received upon the triangle AOB is constant, the curve which is always touched by AB will be an hyperbola whose equation referred to the axes OA, OB is

$$(y - mx)(x - my) = mc^2,$$

where $OC = c$, and m is a constant quantity.

402. If $u = f(x, y, z)$ be a rational function of x, y, z, and if $u = 0$ be the equation to a surface, for a point (a, b, c) of which all the partial differential coefficients of u as far as those of the $(n - 1)^{\text{th}}$ order vanish, shew that the conical surface whose equation is

$$\left\{ (x - a)\frac{d}{da} + (y - b)\frac{d}{db} + (z - c)\frac{d}{dc} \right\}^n f(a, b, c) = 0$$

will touch the proposed surface at the point (a, b, c).

403. Determine the condition to which the vertices of a system of cones which envelope an ellipsoid must be subject, in order that the centres of the ellipses of contact may be equidistant from the centre of the ellipsoid.

404. A plane passes through the vertex of the elliptic paraboloid $2z = \dfrac{x^2}{a} + \dfrac{y^2}{b}$: shew that if the radius of curvature of all its sections of the surface at the vertex be equal to c, the normal to the plane at the origin will lie upon the surface

$$(x^2 + y^2)^3 = c^2 (x^2 + y^2 + z^2) \left(\frac{x^2}{b} + \frac{y^2}{a} \right)^2.$$

405. Find the position in which an ellipsoid must be placed in order that its orthogonal projection may be circular, assuming that the plane on which it is projected must be parallel to its mean axis.

406. Find the equation to the surface generated by a straight line of given length, which moves parallel to the plane of (x, y), with one end in the plane of (y, z) and the other on a given curve, $x = \phi(z)$, in the plane of (x, z).

407. Find the shortest distance between the diagonal of a cube and an edge which it does not meet.

408. Find the equation to the plane which passes through the origin and contains the straight line determined in Example 45.

409. Find the equations to the perpendicular from the origin on the straight line determined in Example 45, and the co-ordinates of the point of intersection.

410. A cube is cut by a plane perpendicular to one of its diagonals: determine the perimeter and area of the section, and the greatest value of the area.

411. Find the point from which the sum of the squares of the perpendiculars on four given planes is a minimum.

412. If one of the diagonals of a rectangular parallelepiped be so placed as to subtend at the eye a right angle, each of the others will at the same time appear under a right angle.

413. Find the largest parallelepiped which can .be inscribed in a given ellipsoid.

414. A series of planes described according to a given law cut an ellipsoid whose equation is

$$\frac{x^2}{a^2} + \frac{y^2}{b^2} + \frac{z^2}{c^2} = 1:$$

find the locus of the centres of all the plane sections.

Example. If the normal to the cutting plane be a generating line of the cone $a^2x^2 + b^2y^2 - c^2z^2 = 0$, the locus will be another cone whose equation is

$$\frac{x^2}{a^2} + \frac{y^2}{b^2} - \frac{z^2}{c^2} = 0.$$

415. Find the condition that the plane $lx + my + nz = p$ may cut the surface $\frac{x^2}{a^2} + \frac{y^2}{b^2} - \frac{z^2}{c^2} = 1$ in two straight lines.

416. A cone whose vertical angle is $90°$ intersects the sphere which touches the axis of the cone at the vertex ; find the projections of the curve of intersection on the plane perpendicular to the axis of the cone, on that perpendicular to the diameter of the sphere at the vertex, and on that which contains these two straight lines: and compare the areas of the latter two.

417. If a section be taken through the axis of an oblate spheroid, and through the directrix of the curve thus formed any plane be drawn cutting the surface, the cone which has for its base the section of the surface by this plane and for its vertex the focus corresponding to the directrix will be one of revolution.

418. Find the equation to the locus of a straight line always intersecting and perpendicular to the straight line $x + y = 0$, $z = 0$, and passing through the perimeter of the parabola $x^2 = lz$, $y = 0$.

T. A. G. 5

419.   All sections of the surface

$$\frac{x^2}{a^2} + \frac{y^2}{b^2} + \frac{z^2}{c^2} = 1,$$

which are at the same distance p from the origin, have their centres on the surface of which the equation is

$$\left(\frac{x^2}{a^2} + \frac{y^2}{b^2} + \frac{z^2}{c^2}\right)^2 = p^2 \left(\frac{x^2}{a^4} + \frac{y^2}{b^4} + \frac{z^2}{c^4}\right).$$

420.   Determine the surface represented by $z^2 = xy + x^2$.

421.   Determine the surface represented by
$$x^2 - y^2 - z^2 + 2yz + x + y - z = 0.$$

422.   The ellipsoid $\dfrac{x^2}{a^2} + \dfrac{y^2}{b^2} + \dfrac{z^2}{c^2} = 1$ is cut by the plane $lx + my + nz = p$: find the equation to an ellipsoid similar to the original ellipsoid and similarly situated, which has its centre at the centre of the plane section and passes through the curve of intersection. Apply the result to find the area of the plane section.

423.   Suppose u a homogeneous function of the second degree, v_1 and v_2 homogeneous functions of the first degree, c_1 and c_2 constants ; then the cone which has its vertex at the origin and for directrix the intersection of $u + v_1 + c_1 = 0$ and $v_2 + c_2 = 0$ is determined by

$$c_2^2 u - c_2 v_1 v_2 + c_1 v_2^2 = 0.$$

424.   Find the area of the section of the cone

$$\frac{x^2}{a^2} + \frac{y^2}{b^2} - \frac{z^2}{c^2} = 0$$

made by the plane $lx + my + nz = p$.

425.   Find the volume contained between the plane and the cone in the preceding Example.

426.   Shew that the tangent plane at any point of an

hyperboloid of two sheets cuts off a constant volume from the asymptotic cone.

427. A straight line passing through a fixed point and having the sum of its inclinations to two fixed straight lines through the same point constant generates a cone of the second order.

Any section perpendicular to either of the fixed straight lines has for its focus its intersection with the fixed straight line.

428. If two surfaces of the second order touch at two points they will in general intersect in two planes.

429. If PQ, PR be any two chords of an ellipsoid in the same plane with a fixed chord PK and inclined at the same angle to it, then the locus of all the possible intersections of QR, $Q'R'$, &c. will be the straight line PK and another straight line.

430. A sphere touches an elliptic paraboloid at the vertex and has its diameter a mean proportional between the parameters of the principal sections of the paraboloid : find the equation to the projection of the curve of intersection on the tangent plane at the vertex.

431. A cone has its vertex at the centre of an ellipsoid and for its directrix a plane section of the ellipsoid : if the cone cut the tangent plane to the ellipsoid at the extremity of one of its axes in a circle, the plane section of the cone and ellipsoid must pass through one of two fixed points in the smaller of the other two axes.

432. C is the centre of an ellipsoid, P an external point which is the vertex of a cone enveloping the ellipsoid, and with any point Q as vertex a cone is constructed having its generating lines parallel respectively to those of this enveloping cone : shew that the cone having its vertex at Q cannot cut the ellipsoid in a plane curve unless Q be on the straight line CP or CP produced.

5—2

433. An oblate spheroid revolves about any diameter: find the equation to the surface which envelops it in every position.

434. If any number of straight lines, drawn in any directions from one given point have their lengths proportional to the cosines of the angles which they severally make with the longest line a, and spheres be described on each of these straight lines as diameters, these spheres will be enveloped by the surface generated by the revolution of the curve

$$r = a \cos^2 \frac{\theta}{2},$$

the given point being the pole.

435. A plane is moved so as to cut off from the co-ordinate planes areas whose sum is always equal to $2n^2$: shew that the surface to which the plane is always a tangent plane is represented by the equation

$$(3x - u)(3y - u)(3z - u) + 9n^2 u = 0,$$

where $u = x + y + z \pm \sqrt{(3n^2 + x^2 + y^2 + z^2 - xy - yz - zx)}$.

436. Find the general functional equation to the surfaces of Example 204.

437. Tangent planes to the surface determined by

$$\frac{x^3}{a^3} + \frac{y^3}{b^3} + \frac{z^3}{c^3} = 1$$

pass through a point P: shew that a sphere can be described through the curve of contact provided P lie on a certain straight line passing through the origin.

438. A straight line BC of given length moves with its extremities always on two fixed straight lines AB, AC at right angles to each other; from the point A a perpendicular AD is drawn to BC, and with radius AD and centre D a circle is described having its plane perpendicular to BC: find the equation to the surface generated by the circle.

439. Determine the surface represented by

$$z^2 = \{\sqrt{(x^2 + y^2)} - a\} \{b - \sqrt{(x^2 + y^2)}\}.$$

440. If r be the radius vector, p the perpendicular from the origin on the tangent, s the arc of the curve,

$$p\frac{dp}{dr} = \frac{p^2}{r} - r^2\frac{d^2r}{ds^2}.$$

441. If ρ be the radius of absolute curvature of a curve, ρ' the distance of the centre of absolute curvature from the origin, r the radius vector, p the perpendicular from the origin on the tangent,

$$r\frac{dr}{dp} = \frac{2\rho^2 p}{r^2 + \rho^2 - \rho'^2}.$$

442. In the surface

$$2cz^2 + z\left(x^2 + 2xy + y^2 - 2c^2\right) - 4cxy = 0$$

find the radius of curvature of a normal section at any point in the axis of x, the tangent to the curve of section at the point being perpendicular to the axis of x.

443. In the surface of the preceding Example find the principal radii of curvature for any point in the axis of x.

444. Assuming that the curvature at any point of a surface is measured by the product of the reciprocals of the principal radii, shew that the lines of equal curvature on the surface of an elliptic paraboloid are projected on a plane perpendicular to its axis in similar ellipses, whose axes are proportional to the parameters of the principal sections of the paraboloid.

445. Conceive a system of surfaces of such a nature that a normal at any point of one of them is a normal to all the rest; let $u = 0$ be the equation to one of these surfaces, let the normal at any point P be the axis of z, and let the distance of the origin from the point P be an harmonic mean between the greatest and least radii of curvature at that point: then shew that at the point P

$$\frac{d^2u}{dx^2} + \frac{d^2u}{dy^2} + \frac{d^2u}{dz^2} = \frac{1}{z}\frac{d^2(uz)}{dz^2}.$$

446. A plane passes through a fixed point; find its position when the volume included between it and the co-ordinate planes is a minimum.

447. A solid angle is formed by three planes; the tangent plane to a given surface meets them, and so forms the base of a tetrahedron. Shew that in general when this tetrahedron is a maximum or a minimum the point of contact is at the centre of gravity of the base of the tetrahedron.

448. The number of normals to a surface of the nth degree which can pass through a given point cannot exceed

$$n^3 - n^2 + n.$$

449. In a surface of the second degree the locus of the points for which the sum of the squares of the normals to the surface is a constant quantity, is a surface of the second degree concentric with the given surface, and having the direction of its principal axes the same.

450. Two spheres being given in magnitude and position, every sphere which intersects them in given angles will touch two other fixed spheres, or cut another at right angles.

451. From the origin are drawn three equal straight lines of length p, such that the inclinations of the first to the axes of x, y, z, are the same as those of the second to y, z, x, and of the third to z, x, y, respectively; three planes are drawn perpendicular to them through their extremities: find the co-ordinates of their common point.

452. A plane cuts the six edges of a tetrahedron at six points; six other points are taken, one in each edge, so as to cut it harmonically: shew that the six planes through these points and the opposite edges of the tetrahedron intersect at one point.

453. *OL, OM, ON* are conjugate semi-diameters in an ellipsoid; a perpendicular is drawn from O on the plane

LMN meeting it at Q; and a diametral plane is drawn parallel to the plane LMN. Shew that the cone which has its vertex at Q, and for its base the section of the ellipsoid by the diametral plane, is of constant volume.

454. Shew that the foci of all parabolic sections of the surface

$$\frac{y^2}{a} + \frac{z^2}{b} = x,$$

lie on the surface

$$\left(x - \frac{y^2}{a} - \frac{z^2}{b}\right)\left(\frac{y^2}{a} + \frac{z^2}{b}\right) = \frac{ab}{4}\left(\frac{y^2}{a^2} + \frac{z^2}{b^2}\right).$$

455. If normals be drawn to the ellipsoid

$$\frac{x^2}{a^2} + \frac{y^2}{b^2} + \frac{z^2}{c^2} = 1$$

at the points where it is cut by the cone

$$\frac{l}{x} + \frac{m}{y} + \frac{n}{z} = 0,$$

prove that all these normals will pass through a diameter of the ellipsoid.

456. Shew that a straight line parallel to the least axis of an ellipsoid will be the directrix of two plane sections of the ellipsoid, provided the straight line be situated between two definite cylindrical surfaces.

457. Investigate an equation to the surface generated by a straight line which always meets three straight lines, which do not pass through the same point and are not all parallel to the same plane.

Two surfaces are so generated, the three fixed straight lines for one surface being parallel to the three fixed straight lines for the other: shew that if the two surfaces intersect on two planes, one plane will be parallel to four of the fixed straight lines, and the other plane will be parallel to the other two.

458.   A cone whose vertex is any point of the hyperbola

$$x = 0, \quad \frac{z^2}{k^2} - \frac{y^2}{h^2} = 1,$$

envelopes the ellipsoid

$$\frac{x^2}{a^2} + \frac{y^2}{b^2} + \frac{z^2}{c^2} = 1,$$

whose least semi-axis is c; and h and k satisfy the relation

$$b^2 - c^2 = \frac{b^2(a^2 - b^2)}{h^2} + \frac{c^2(a^2 - c^2)}{k^2}:$$

shew that the directrices of all the sections of the ellipsoid made by the planes of contact lie in one or other of two fixed planes.

459.   Prove that the foci of all centric sections of the surface

$$ax^2 + by^2 + cz^2 = 1$$

lie on the surface

$$\frac{(x^2 + y^2 + z^2)(1 - ax^2 - by^2 - cz^2)}{ax^2 + by^2 + cz^2}$$

$$= \frac{(c - b)^2 y^2 z^2 + (a - c)^2 z^2 x^2 + (b - a)^2 x^2 y^2}{a(c - b)^2 y^2 z^2 + b(a - c)^2 z^2 x^2 + c(b - a)^2 x^2 y^2}.$$

460.   The equations to points in a curve are

$$x = 4a \cos^3 \theta, \quad y = 4a \sin^3 \theta, \quad z = 3c \cos 2\theta:$$

find the equations of the normal and osculating planes; and determine the relation between c and a that the locus of the centre of the sphere of curvature may be a curve similar to the original curve.

RESULTS OF THE EXAMPLES.

1. $\dfrac{x}{a} + \dfrac{y}{b} + \dfrac{z}{c} = 2.$

2. $\dfrac{x}{a} + \dfrac{y}{b} - \dfrac{z}{c} = 1.$

3. $\dfrac{y}{b} = \dfrac{z}{c}; \quad \dfrac{x}{a} = \dfrac{z}{c}.$

4. $\dfrac{x}{a} = \dfrac{y}{b} = \dfrac{z}{c}.$

5. $\dfrac{x}{a} = \dfrac{y-b}{-b} = \dfrac{z}{c}; \quad \dfrac{x-a}{-a} = \dfrac{y}{b} = \dfrac{z}{c}.$

6. $\dfrac{2abc}{\sqrt{(a^2b^2 + b^2c^2 + c^2a^2)}}.$

7. $\dfrac{2abc}{\sqrt{(a^2b^2 + b^2c^2 + c^2a^2)}}.$

8. $\cos^{1} \dfrac{ab}{\sqrt{(a^2 + c^2)}\,\sqrt{(b^2 + c^2)}}.$

9. $\cos^{-1} \dfrac{3}{\sqrt{(a^2 + b^2 + c^2)}\,\sqrt{\left(\dfrac{1}{a^2} + \dfrac{1}{b^2} + \dfrac{1}{c^2}\right)}}.$

10. $\cos^{-1} \dfrac{a^2 + b^2 - c^2}{a^2 + b^2 + c^2}.$

11. $\dfrac{x}{a} = \dfrac{2y}{b} = \dfrac{2z}{c}.$

12. The equation may be written
$$(x \cos \beta - y \cos \alpha)^2 + (y \cos \gamma - z \cos \beta)^2 + (z \cos \alpha - x \cos \gamma)^2 = 0\,;$$
thus it represents the straight line
$$\frac{x}{\cos \alpha} = \frac{y}{\cos \beta} = \frac{z}{\cos \gamma}.$$

13. $30^{\circ}.$

14. $\dfrac{Ax + By + Cz - D}{Aa + B\beta + C\gamma - D} = \dfrac{A'x + B'y + C'z - D'}{A'a + B'\beta + C'\gamma - D'}.$

15. $D'(Ax + By + Cz) = D(A'x + B'y + C'z)$; the condition is
$$\left(\frac{D}{D'}\right)^2 = \frac{A^2 + B^2 + C^2}{A'^2 + B'^2 + C'^2}.$$

16. $(x - a)\{m(c - c') - n(b - b')\} + \&c. = 0.$

17.   First obtain the equation to the plane which contains the two given straight lines; this is

$$x \left(m_1 n_2 - m_2 n_1\right) + y \left(n_1 l_2 - n_2 l_1\right) + z \left(l_1 m_2 - l_2 m_1\right) = 0 ;$$

then find the equation to a plane which is perpendicular to the plane just determined and equally inclined to the two given straight lines.

21. $\frac{5}{3} \sqrt{2}$.   22. $(x - a) \left(mn' - m'n\right) + \&c. = 0$.

23.   The question is indeterminate, since an unlimited number of straight lines can be drawn as required.

24.   Suppose the given plane to be determined by

$$Ax + By + Cz = D ;$$

and suppose that the line of intersection is to lie in the plane of (x, y); then we may assume for the equation to the required plane $Ax + By + \lambda z = D$, and determine λ suitably.

27. 60°.   28.   Four straight lines.

29.   The condition is $a^2 + b^2 + c^2 + 2abc = 1$.

30. $\left(p_1 \cos a - p \cos a_1\right) x + \left(p_1 \cos \beta - p \cos \beta_1\right) y + \lambda z = 0$,
where $\left(p_1 \cos a - p \cos a_1\right)^2 + \left(p_1 \cos \beta - p \cos \beta_1\right)^2 + \lambda \left(p_1 \cos \gamma - p \cos \gamma_1\right) = 0$.

31.   Let the given point be (a, b, c) and the equations to the given planes

$$Ax + By + Cz = D, \quad A'x + B'y + C'z = D' ;$$

the required equation is

$$(x - a) \left(BC' - B'C\right) + \&c. = 0.$$

37. $\dfrac{p + p'}{\sqrt{(2 + 2k)}}$ and $\dfrac{p - p'}{\sqrt{(2 - 2k)}}$, where $k = ll' + mm' + nn'$.

41.   Let r denote the distance between (a, β, γ) and (x, y, z); then

$$\left(A^2 + B^2 + C^2\right) r^2 = \left(A^2 + B^2 + C^2\right) \left\{(x - a)^2 + (y - \beta)^2 + (z - \gamma)^2\right\}$$
$$= \left\{B (z - \gamma) - C (y - \beta)\right\}^2 + \left\{C (x - a) - A (z - \gamma)\right\}^2 + \left\{A (y - \beta) - B (x - a)\right\}^2$$
$$+ \left\{A (x - a) + B (y - \beta) + C (z - \gamma)\right\}^2.$$

The last term is *constant* if (x, y, z) be in the given plane; hence the *least* value of r^2 is obtained by making the other three terms vanish.

44. The exceptional case is when the line of intersection of two of the planes is perpendicular to the third plane.

45. $\dfrac{1}{a}\left\{x-\dfrac{b\gamma-c\beta}{a^2+b^2+c^2}\right\}=\dfrac{1}{b}\left\{y-\dfrac{ca-a\gamma}{a^2+b^2+c^2}\right\}=\dfrac{1}{c}\left\{z-\dfrac{a\beta-ba}{a^2+b^2+c^2}\right\}.$

46. The condition is $al+bm+cn=0$; then the equations may be written

$\dfrac{1}{l}\left\{x-\dfrac{cm-bn}{l^2+m^2+n^2}\right\}=\&\mathrm{c.}$; or thus, $\dfrac{1}{l}\left\{x-\dfrac{cn-bm}{3mn}\right\}=\&\mathrm{c.}$;

or thus, $\dfrac{1}{l}\left\{x-\dfrac{c-b}{l+m+n}\right\}=\&\mathrm{c.}$

47. The equations to the perpendicular are

$$\frac{x}{mc-nb}=\frac{y}{na-lc}=\frac{z}{lb-ma}.$$

49. There are two such straight lines determined by the given equation $lx+my+nz=0$, combined with

$$x\sqrt{(l^2+n^2)}=\pm y\sqrt{(m^2+n^2)}.$$

50. Let x, y, z be the co-ordinates of any point in the straight line; let l, m, n be proportional to its direction cosines; then it may be shewn that

$$\cos^2 a=\frac{(nx-lz)^2(mx-ly)^2}{\{(nx-lz)^2+(ny-mz)^2\}\{(mx-ly)^2+(mz-ny)^2\}};$$

from this we may shew that $\tan^2 a=A(mz-ny)^4$ where A is a symmetrical function of x, y, z, and l, m, n.

52. $\dfrac{x}{a^2}=\dfrac{y}{b^2}=\dfrac{z}{c^2}=\dfrac{1}{\sqrt{(a^2+b^2+c^2)}}$; $\dfrac{x}{a}=\dfrac{y}{b}=\dfrac{z}{c}=\dfrac{1}{\sqrt{3}}$.

53. Here $e=\dfrac{\sqrt{(a^2-b^2)}}{a}$, $e'=\dfrac{\sqrt{(b^2-c^2)}}{b}$.   54. A sphere.

55. The polar equation to the locus is

$$a^2b^2=(r^2+c^2)(a^2\sin^2\theta+b^2\cos^2\theta).$$

56. The straight line $\dfrac{x}{l}=\dfrac{y}{m}=\dfrac{z}{n}$.

58. The given equation may be written thus, $r^2=(l^2+m^2+n^2)p^2$ where r is the distance of the point (x,y,z) from the origin, and p is

the perpendicular from the point (x, y, z) on the plane $lx+my+nz=0$. If $l^2 + m^2 + n^2$ is greater than unity the locus is a right circular cone; the cosine of the semivertical angle is $\dfrac{1}{\sqrt{(l^2 + m^2 + n^2)}}$. If $l^2 + m^2 + n^2$ is equal to unity the locus is a straight line; see Example 12. If $l^2 + m^2 + n^2$ is less than unity the locus is a point, namely the origin.

59.   The locus is a sphere if C be an acute angle, a point if C be a right angle, and impossible if C be an obtuse angle.

60. $k^2 = \dfrac{a^2}{a^2 - b^2}$, $m^2 = \dfrac{a^2(b^2 - c^2)}{c^2(a^2 - b^2)}$.

62.   The equations to the cones are
$$\frac{(x - a)^2}{a^2} = \frac{y^2}{b^2} + \frac{z^2}{c^2}, \quad \frac{(y - b)^2}{b^2} = \frac{z^2}{c^2} + \frac{x^2}{a^2}.$$

63.   Let (a, β, γ) be the external point; then the required equation is
$$\frac{x^2}{a^2} + \frac{y^2}{b^2} + \frac{z^2}{c^2} = \left(\frac{xa}{a^2} + \frac{y\beta}{b^2} + \frac{z\gamma}{c^2}\right)^2.$$

66.   Let (a, β, γ) be the given point; then the required equation is
$$a^2(b^2 - c^2)\frac{x}{a} + b^2(c^2 - a^2)\frac{y}{\beta} + c^2(a^2 - b^2)\frac{z}{\gamma} - 0.$$

67.   A right circular cone.   70.   Take for the equation to the ellipsoid $\dfrac{x^2}{a^2} + \dfrac{y^2}{b^2} + \dfrac{z^2}{c^2} - 1$, and for the equation to the sphere $(x - h)^2 + y^2 + z^2 = r^2$; then the equation to the cylinder will be
$$(x - h)^2 + y^2 - \frac{c^2 x^2}{a^2} - \frac{c^2 y^2}{b^4} = r^2 - c^2,$$

71.   The circle circumscribed about the opposite face may be determined by the equations
$$\frac{x}{a} + \frac{y}{b} + \frac{z}{c} = 1, \; x^2 + y^2 + z^2 = ax + by + cz.$$

A *subcontrary* section is a circular section. The equation to the cone may be written thus
$$(ax + by + cz - k)\left(\frac{x}{a} + \frac{y}{b} + \frac{z}{c}\right) = x^2 + y^2 + z^2 - k\left(\frac{x}{a} + \frac{y}{b} + \frac{z}{c}\right);$$

hence we see that the plane $ax + by + cz - k = 0$ cuts the cone in a circle. If the vertex of the cone be at the angular point which is at the other end of the edge c, the equation to the cone is

$$x^2 + y^2 = (ax + by)\left(1 - \frac{z}{c}\right);$$

and the plane $ax + by - cz = 0$ is parallel to the subcontrary sections.   74. An hyperbolic paraboloid.

75. The locus is determined by the equation to the ellipsoid combined with $xyz = $ constant.

76. Suppose the fixed point in the axis of z; then we have to make xy a maximum while $\frac{x^2}{a^2} + \frac{y^2}{b^2} = $ a constant.

85. Circles.   88. Let $\dfrac{x' - x}{l} = \dfrac{y' - y}{m} = \dfrac{z' - z}{n} = r$ be the equations to a straight line which passes through the point (x, y, z) and coincides with the given surface. Substitute the values of x', y', z' in the given equation; we thus obtain a quadratic in r which ought to be *identically* true. This leads to

$$\frac{l^2}{a^2} + \frac{m^2}{b^2} - \frac{n^2}{c^2} = 0, \quad \frac{lx}{a^2} + \frac{my}{b^2} - \frac{nz}{c^2} = 0.$$

Eliminate m from these two equations; thus we obtain a quadratic in $\dfrac{l}{n}$, and the product of the roots becomes known. The result may be written thus

$$\frac{l_1 l_2}{n_1 n_2} = \frac{x^2 - a^2}{z^2 + c^2}.$$

$$\text{Similarly } \frac{m_1 m_2}{n_1 n_2} = \frac{y^2 - b^2}{z^2 + c^2}.$$

$$\text{Hence } 1 + \frac{l_1 l_2}{n_1 n_2} + \frac{m_1 m_2}{n_1 n_2} = \frac{z^2 + c^2 + x^2 - a^2 + y^2 - b^2}{z^2 + c^2}.$$

92. The condition is $amn + bnl + clm \quad 0.$

93. When the given conditions hold, the equation becomes
$(b'c'x + c'a y + a'b'z)^2 + 2a'b'c' (a''x + b''y + c''z) - a'b'c'f = 0;$
this represents a parabolic cylinder.

95.   Either $b' = 0$ and $c'^2 = (a - c)(b - c)$,

  or $c' = 0$ and $b'^2 = (a - b)(c - b)$.

96.   An hyperbola.

97.   The semi-axes are the positive values of r found from

$$\frac{a^2 l^2}{r^2 - a^2} + \frac{b^2 m^2}{r^2 - b^2} + \frac{c^2 n^2}{r^2 - c^2} = 0.$$

102.   If an ellipsoid be cut by a plane through its centre whose direction cosines are l, m, n, the area of the section is known to be $\dfrac{\pi abc}{\sqrt{(l^2 a^2 + m^2 b^2 + n^2 c^2)}}$.   Now if we seek the maximum and minimum values of $l^2 a^2 + m^2 b^2 + n^2 c^2$, with the condition that the plane is to contain the straight line $\dfrac{x}{\lambda} = \dfrac{y}{\mu} = \dfrac{z}{\nu}$, we obtain this quadratic

$$\frac{\lambda^2}{u^2 - a^2} + \frac{\mu^2}{u^2 - b^2} + \frac{\nu^2}{u^2 - c^2} = 0;$$

thus the product of the maximum and minimum values is

$$a^2 b^2 c^2 \left(\frac{\lambda^2}{a^2} + \frac{\mu^2}{b^2} + \frac{\nu^2}{c^2} \right).$$

103.   If a, β, γ be co-ordinates of the centre of the section it will be found that

$$\frac{a}{a^2 l} = \frac{\beta}{b^2 m} = \frac{\gamma}{c^2 n} = \frac{\delta}{a^2 l^2 + b^2 m^2 + c^2 n^2}.$$

The final result is

$$\frac{(a^2 l^2 + b^2 m^2 + c^2 n^2 - \delta^2)\, \pi abc}{(a^2 l^2 + b^2 m^2 + c^2 n^2)^{\frac{3}{2}}}.$$

See also *Differential Calculus*, Chapter XVI., where the solution is fully worked out.

104.   The first part may be shewn by taking the general equation

$$ax^2 + by^2 + cz^2 + 2a'yz + 2b'zx + 2c'xy + 2a''x + 2b''y + 2c''z + f = 0;$$

and the cutting planes may be supposed parallel to that of (x, y). For the latter part we must find the ratio of similarly situated lines in the two sections, and the square of this ratio will be the ratio of the areas. 105. This is a mode of expressing the result of the two preceding Examples. The result may also be obtained very easily by referring the ellipsoid to conjugate diameters, two of which are in a plane parallel to the planes considered.

106. The equation to the ellipsoid is

$$\frac{x^2}{a^2} + \frac{y^2}{b^2} + \frac{z^2}{c^2} = \frac{x\alpha}{a^2} + \frac{y\beta}{b^2} + \frac{z\gamma}{c^2},$$

where (α, β, γ) is the external point.

109. Hyperboloid of one sheet. 110. Hyperboloid of one sheet. 111. Hyperbolic paraboloid.

112. Parabolic cylinder. 113. Hyperboloid of two sheets if f be positive. 114. Right circular cylinder if f be positive.

115. Hyperboloid of revolution of two sheets if f be greater than 48, cone if $f - 48$, hyperboloid of revolution of one sheet if f be less than 48.

116. Hyperboloid of revolution of two sheets.

117. $\dfrac{2c\gamma}{3\beta}$, $\dfrac{2c\gamma}{3a}$, 0 are the co-ordinates.

118. There is a line of centres, given by the equations
$$k = 0, \quad h - l + 1 = 0.$$

119. $-\frac{1}{2}$, 0, 0 are the co-ordinates.

122. A surface of revolution, the axis of which is $x = y = z$.

124. Use the equation in Example 65; the locus is a sphere, the radius of which is $\sqrt{(a^2 + b^2 + c^2 + h^2 + k^2)}$.

133. The sine of the angle is $\dfrac{a^2 - c^2}{a^2 + c^2}$ supposing a, b, c in descending order of magnitude. 142. See a similar example in Geometry of Two Dimensions, *Plane Co-ordinate Geometry*, Art. 286. 144. Refer the ellipsoid to conjugate diameters such that the plane containing two of them is parallel to the cutting plane.

157.　$b'^2 - ac$ must be positive and $(b'^2 - ac)(a'^2 - bc) = (a'b' - cc')^2$.

159.　Let　$x \cos \theta_1 + y \cos \theta_2 + z \cos \theta_3 = u,$
　　　　　　　$x \sin \theta_1 + y \sin \theta_2 + z \sin \theta_3 = v \,;$

then the second equation is $uv - 0$, and the first is $(u+v)(u-v) = 0$.
The line of intersection will be found to be determined by

$$\frac{x}{\sin(\theta_3 - \theta_2)} = \frac{y}{\sin(\theta_1 - \theta_3)} = \frac{z}{\sin(\theta_2 - \theta_1)}.$$

160.　The other straight lines are $x = 0,\ yb - zc = 0\,;\ y = 0,$
$zc - xa = 0\,;\ z = 0,\ xa - yb = 0.$

161.　The cosine of the inclination of the plane to the axis.

170.　Take the equation to an enveloping cylinder

$$\left(\frac{l^2}{a^2} + \frac{m^2}{b^2} + \frac{n^2}{c^2}\right)\left(\frac{x^2}{a^2} + \frac{y^2}{b^2} + \frac{z^2}{c^2} - 1\right) = \left(\frac{lx}{a^2} + \frac{my}{b^2} + \frac{nz}{c^2}\right)^2,$$

and apply the tests in order that it may be a surface of revolution.

182.　$\phi(x, y, z) = \phi(a + x - a,\ \beta + y - \beta,\ \gamma + z - \gamma)$

$$= u + (x - a)\frac{du}{da} + (y - \beta)\frac{du}{d\beta} + (z - \gamma)\frac{du}{d\gamma}$$

$$+ \frac{(x - a)^2}{2}\frac{d^2u}{da^2} + \frac{(y - \beta)^2}{2}\frac{d^2u}{d\beta^2} + \frac{(z - \gamma)^2}{2}\frac{d^2u}{d\gamma^2}$$

$$+ (y - \beta)(z - \gamma)\frac{d^2u}{d\beta\,d\gamma} + (z - \gamma)(x - a)\frac{d^2u}{d\gamma\,da} + (x - a)(y - \beta)\frac{d^2u}{da\,d\beta}.$$

The equation to the tangent plane to the surface $\phi(x, y, z) = 0$
at the point (x, y, z) is

$$(x' - x)\frac{d\phi}{dx} + (y' - y)\frac{d\phi}{dy} + (z' - z)\frac{d\phi}{dz} = 0\,;$$

this may be written

$$(x' - x)\left\{\frac{du}{da} + (x - a)\frac{d^2u}{da^2} + (z - \gamma)\frac{d^2u}{d\gamma\,da} + (y - \beta)\frac{d^2u}{da\,d\beta}\right\} + \&c. = 0.$$

Now suppose this plane to pass through the point (a, β, γ), and
we obtain a relation which by means of the equation $\phi(x, y, z) = 0$,
transformed as above, reduces to

$$2u + (x - a)\frac{du}{da} + (y - \beta)\frac{du}{d\beta} + (z - \gamma)\frac{du}{d\gamma} = 0.$$

183. Let a straight line pass through the point (α, β, γ) and through the point (x, y, z) on the surface; and let (x', y', z') be any other point on this straight line; then we have

$$\frac{x-\alpha}{x'-\alpha} = \frac{y-\beta}{y'-\beta} = \frac{z-\gamma}{z'-\gamma} = r \text{ say; therefore}$$

$$x = \alpha + r(x'-\alpha), \quad y = \beta + r(y'-\beta), \quad z = \gamma + r(z'-\gamma).$$

Substitute these values of x, y, z in $\phi(x, y, z) = 0$; we thus obtain a *quadratic* in r corresponding to the two points at which the straight line through (α, β, γ) and (x', y', z') meets the surface. The condition that this quadratic should have *equal* roots leads to the equation to the required cone, x', y', z' being the variable co-ordinates.

185. $a'yz + b'zx + c'xy = 0.$

195. $\dfrac{a^2 x^2}{a^2 - k^2} + \dfrac{b^2 y^2}{b^2 - k^2} + \dfrac{c^2 z^2}{c^2 - k^2} = 0.$

196. The surface is determined by the equation

$$y^2 - 2ay + \frac{z^2 - 2kaz}{k^2} + a^2 - 0,$$

where k is a known constant, and $a = x \left\{ 1 + \sqrt{\left(\dfrac{2x}{kc}\right)} + \dfrac{x}{kc} \right\}.$

200. $\dfrac{x^2}{a^2} + \dfrac{y^2}{b^2} \quad \left(1 - \dfrac{z^2}{c^2}\right)\left(\dfrac{x}{a} + \dfrac{y}{b}\right)^2.$

201. $\left(\dfrac{x}{a} \pm \dfrac{y}{b}\right)^2 + \dfrac{z^2}{c^2} \quad 1.$

205. Take for the equations to the straight lines those in Example 74; let a be the radius of the circle; the required equation is

$$(mxc^2 - cyz)^2 + m^2(mcxz - c^2 y)^2 - m^2 a^2 (c^2 - z^2)^2.$$

212. (1) is not a developable surface; (2 is.

214. $z + \sqrt{(x^2 + y^2 + z^2)} = \phi\left(\dfrac{y}{x}\right).$

T. A. G.

6

215. $z^{n+1} - k^{n+1} - x^{n+1} \phi \left(\dfrac{y}{x} \right)$.   216. $\dfrac{x^2 + y^2}{a^2} = \dfrac{(z - c)^2}{c^2}$.

218. $(ax + by + cz)^2 = r^2 \{ (x + a)^2 + (y + b)^2 + (z + c)^2 \}$.

219. Let the equations to the parabolas be

$$\left. \begin{array}{c} y - 0 \\ z = 4ax \end{array} \right\}, \text{ and } \left. \begin{array}{c} z = 0 \\ y^2 = 4a'x \end{array} \right\};$$

the equation to the surface is $4x - \left\{ \dfrac{y}{\sqrt{a'}} + \dfrac{z}{\sqrt{a}} \right\}^2 = 0$.

224. $(x + a) \left\{ \dfrac{y^2}{a + x - b} + \dfrac{z^2}{a + x - c} + x - a \right\} = 0$,

where $2x = \dfrac{y^2}{b} + \dfrac{z^2}{c}$ is the equation to the paraboloid.

225. $(x^2 + \mu^2 y^2)^2 = \dfrac{k^6}{z^2}$, where μ and k are known constants.

227. Use conjugate diameters as axes, two of them being parallel to the plane of the circles, and the third passing through the centres of the circular sections.

230. $y = \pm a \cos \beta$.   231. $\dfrac{x'x}{a^2} + \dfrac{y'y}{b^2} = \dfrac{1}{2}, \ z = \dfrac{c}{\sqrt{2}}$.

232. $x' - x = \dfrac{2y}{a - 2x} (y' - y) = - \dfrac{2z}{a} (z' - z)$.

233. The curve is determined by the given equation combined with $4z^2 = x^2 + y^2$.

239. They are proportional to $\dfrac{d^2x}{ds^2}, \ \dfrac{d^2y}{ds^2}$, and $\dfrac{d^2z}{ds^2}$.

240. $c (x'y - y'x) + ab (z' - z) = 0$.   241. A plane passing through the line of intersection of the normal plane at the point (x, y, z) and the consecutive normal plane, and perpendicular to the first normal plane.   243. $\dfrac{(h^2 a^2 + k^2 z^2)^{\frac{3}{2}}}{a^2 h^2 \sqrt{(h^2 + k^2)}}$.   246. A circle.

251. The curves of greatest inclination to the plane of (x, y) are determined by the given equation combined with $x^2 - y^2 =$ constant. 252. The curves are determined by the given equation combined with $x^2 - y^2 =$ constant.

253. $x^2 + 3ay = 0$, $27acz + x^3 - 0$. 257. $27a(x+y)^2 = 16(z-a)^3$.

263. There are two surfaces determined by the equations
$$27a(y + z - 4a)^2 = 4(x - 3a)^3,$$
and
$$27a(y - z)^2 - 4(x - 7a)^3.$$
The specified evolute is determined by the second equation combined with $27a(y + z + 4a)^2 = 2(x - a)^3$.

264. $a + b = A + B$, $c'^2 - ab = C'^2 - AB$.

269. $z = A \tan^{-1} \dfrac{y}{x} + B$. 270. The surface is that formed by the revolution of the catenary $y = \dfrac{c}{2}(e^{\frac{x}{c}} + e^{-\frac{x}{c}})$ round the axis of x.

281.
$$\frac{y + \sqrt{(y^2 + a^2)}}{x + \sqrt{(x^2 + a^2)}} = \text{constant},$$
$$\{y + \sqrt{(y^2 + a^2)}\}\{x + \sqrt{(x^2 + a^2)}\} = \text{constant}.$$

288. $x = 0$, $y = -\sqrt{(a\beta - \beta^2)}$, $z = \dfrac{1}{2}(a - \beta)$; supposing $a > \beta$.

299. The condition is $abc + 2a'b'c' - aa'^2 - bb'^2 - cc'^2 = 0$.

302. $A\alpha + B\beta + C\gamma = 0$, is the equation to the plane which is parallel to the face of which the area is D, and which passes through the opposite vertex.

305. The equation represents a cone which touches the planes represented by $u_1 = 0$, $u_2 = 0$, $u_3 = 0$.

306. The equation represents a cone containing the lines of intersection of the planes $u_1 = 0$, $u_2 = 0$, $u_3 = 0$.

313. An ellipsoid, a point, or an impossible locus according as a is $> =$ or < 0.

314. An hyperboloid of one sheet, a cone, or an hyperboloid of two sheets according as a is $> =$ or < 0.

315. An elliptic cylinder, a straight line, or an impossible locus according as a is $> =$ or < 0.

316. An hyperbolic cylinder, two planes, or an hyperbolic cylinder according as a is $> =$ or < 0.

317. Two parallel planes, one plane, or an impossible locus according as a is $> \doteq$ or < 0.

318. An elliptic paraboloid. 319. An hyperbolic paraboloid. 320. A parabolic cylinder. See for the last eight questions *The Mathematician*, Vol. III. p. 195.

In Examples 321...329 the symbols r, θ, ϕ are the usual polar co-ordinates. 321. A right circular cone having its vertex at the origin and its axis coincident with the axis of z. 322. A plane containing the axis of z. 323. A sphere having its centre at the origin. 324. A series of right circular cones having their vertices at the origin and their axes coincident with the axis of z. 325. A series of planes containing the axis of z. 326. A series of spheres having the origin for centre. 327. A surface of revolution round the axis of z. 328. A surface such that any section made by a plane which contains the axis of z is a circle with the origin for centre. 329. A conical surface generated by straight lines which all pass through the origin.

331. A sphere having the origin on its surface.

337. $\pm \dfrac{abc}{6} \left(\dfrac{x}{a} + \dfrac{y}{b} + \dfrac{z}{c} - 1 \right)$.

340. Let c be a side of the hexagon, a an edge of the cube; then $a = \dfrac{c\sqrt{3}}{\sqrt{2}}$.

342. Let a be the edge of the cube; the height of the luminous point above the given plane is $a\left(2 + \sqrt{2}\right)$.

347. The figure formed by the revolution of an hyperbola round its conjugate axis.

352. $(1 - m^2)(z^2 - c^2) = y^2 - m^2 x^2$; the axes being as in Example 74.

354. The equation to the surface is

$$2cz^2 + z\{(x + y)^2 - 2c^2\} - 4cxy - 0 \; ; \; \text{the volume is } \frac{83\pi c^3}{240\sqrt{2}}.$$

355. $y = z = 0$.   359. $z^2 - 2(x^2 + y^2)$.

360. $\dfrac{D^2}{s^2} = \dfrac{(a - r)(r - b)}{r^2}, \quad c - \dfrac{ab}{a + b - r}.$

363. $x^{a^2} - k y^{b^2}$, and $x^2 + y^2 + z^2 - \log(k' y^{2b^2})$.

380. $(x^2 + y^2 + z^2)^3 - 6c^3 xyz$ where c^3 is the constant volume.

383. $c^2 xyz - a^3(x^2 + y^2)$.   384. Hyperbolic paraboloid.

389. The volume is $\dfrac{\pi a^3}{15}$.

392. $\left(\dfrac{x}{a} + \dfrac{y}{b} + \dfrac{z}{c} - 1\right)^2 - 2\left(\dfrac{x^2}{a^2} + \dfrac{y^2}{b^2} + \dfrac{z^2}{c^2} - 1\right).$

393. The origin; the locus is the axis of z.

398. $4c^2\left(\dfrac{x^2}{a^2} + \dfrac{y^2}{b^2}\right) + x^2 + y^2 - l^2$, where $2l$ is the length of the straight line, and $\dfrac{x^2}{a^2} + \dfrac{y^2}{b^2} = \dfrac{z^2}{c^2}$ the equation to the given cone.

405. The longest axis of the ellipsoid must be inclined to the plane at an angle whose cosine is $\left(\dfrac{b^2 - c^2}{a^2 - c^2}\right)^{\frac{1}{2}}$.

406. $y^2\{\phi(z)\}^2 = \{x - \phi(z)\}^2[c^2 - \{\phi(z)\}^2]$.

408. $x\{a(a^2 + b^2 + c^2) - a(a\alpha + b\beta + c\gamma)\} + \ldots = 0$.

409. $\dfrac{x}{b\gamma - c\beta} = \&c.$; co-ordinates those given in the result of Example 45.   410. Let a be the edge, p the distance of the section from one corner. Then from $p = 0$ to $p = \dfrac{a}{\sqrt{3}}$ the section is

an equilateral triangle; the perimeter is $3p\sqrt{6}$ and the area $\dfrac{p^2 3\sqrt{3}}{2}$.

From $p = \dfrac{a}{\sqrt{3}}$ to $p = \dfrac{2a}{\sqrt{3}}$ the section is a hexagon having three sides equal and also the other three sides equal; the perimeter is $3p\sqrt{6} - 3\sqrt{2}\,(p\sqrt{3} - a)$, that is $3a\sqrt{2}$, and the area is $\dfrac{p^2\,3\sqrt{3}}{2} - 3\{\sqrt{2}\,(p\sqrt{3} - a)\}^2\dfrac{\sqrt{3}}{4}$, that is $9ap - 3p^2\sqrt{3} - \dfrac{3a^2\sqrt{3}}{2}$.

From $p = \dfrac{2a}{\sqrt{3}}$ to $p = a\sqrt{3}$ the results may be obtained from those in the first case by putting $a\sqrt{3} - p$ instead of p. The area is greatest when $p = \dfrac{a\sqrt{3}}{2}$.

415. $p^2 = l^2a^2 + m^2b^2 - n^2c^2.$ 418. $lz = x^2 - y^2.$

421.   Two planes.

422. $\dfrac{(x-a)^2}{a^2} + \dfrac{(y-\beta)^2}{b^2} + \dfrac{(z-\gamma)^2}{c^2} = 1 - \dfrac{p^2}{a^2l^4 + b^2m^2 + c^2n^3}.$

424. $\dfrac{\pi abcp^2}{(l^2a^2 + m^2b^2 - n^2c^2)^{\frac{3}{2}}}.$ 425. $\dfrac{\pi abcp^3}{3\,(l^2a^2 + m^2b^2 - n^2c^2)^{\frac{3}{2}}}.$

426.   Volume is $\dfrac{\pi abc}{3}$.

436. $z = x\phi\,(lx + my + nz) + \psi\,(lx + my + nz).$

438. $z^2 = \dfrac{2axy}{(x^2 + y^2)^{\frac{1}{2}}} - x^2 - y^2.$ 439.   Surface of revolution formed by revolving a circle about a straight line in its plane.

451.   At the common point $x = y = z$.

452.   Take for the faces of the tetrahedron the equations
$$t = 0, \quad u = 0, \quad v = 0, \quad w = 0;$$
and for the equation to the plane cutting the four edges
$$kt + lu + mv + nw = 0.$$
It will be found that the equations to the six planes are respectively $kt = lu$, $kt - mv$, $kt = nw$, $lu = mv$, $lu = nw$, $mv = nw$; thus the common point is determined by
$$kt = lu = mv = nw.$$

453. With the notation of Example 126 the equation to the plane LMN will be found to be

$$\frac{x}{a^2}(x_1 + x_2 + x_3) + \frac{y}{b^2}(y_1 + y_2 + y_3) + \frac{z}{c^2}(z_1 + z_2 + z_3) = 1.$$

The area of the section of the ellipsoid by the diametral plane is known: see the Result of Example 103. The volume will be found to be $\dfrac{\pi abc}{3\sqrt{3}}$.

456. Transform the origin to the point $(-h, -k, 0)$; then the equation to the ellipsoid is

$$\frac{(x'-h)^2}{a^2} + \frac{(y'-k)^2}{b^2} + \frac{z'^2}{c^2} = 1.$$

For a section through the point inclined to the plane (zx) at an angle θ we have

$$\frac{(x''\cos\theta - h)^2}{a^2} + \frac{(x''\sin\theta - k)^2}{b^2} + \frac{z''^2}{c^2} = 1 \quad \dots\dots\dots(1).$$

The equation to an ellipse referred to a directrix and the straight line including the major axis is

$$x'^2(1-e^2) + z''^2 - \frac{2A(1-e^2)}{e}x'' + \frac{A^2(1-e^2)^2}{e^2} = 0 \quad \dots\dots(2).$$

From the comparison of (1) and (2) we deduce

$$\frac{h^2}{a^2} + \frac{k^2}{b^2} - 1 = c^2\left(\frac{h\cos\theta}{a^2} + \frac{k\sin\theta}{b^2}\right)^2.$$

This may be regarded as a quadratic in $\tan\theta$; and to ensure real values it will be found that we must have

$$\frac{h^2}{a^2} + \frac{k^2}{b^2} \text{ greater than } 1,$$

and

$$\frac{h^2}{a^2}\left(1 - \frac{c^2}{a^2}\right) + \frac{k^2}{b^2}\left(1 - \frac{c^2}{b^2}\right) \text{ less than } 1.$$

457. The equation to one surface will be found to be of the form $ayz + bzx + cxy + abc - 0$, say $u - 0$; and the equation to the other surface must be of the form

$$Ayz + Bzx + Cxy + \alpha x + \beta y + \gamma z + \delta = 0, \text{ say } v = 0.$$

If the surfaces meet in plane curves $v - \lambda u = 0$ must be capable of being decomposed into two linear factors.

458.   Let Y and Z be the co-ordinates of the vertex; let $2p$ and $2q$ be the axes of the section, the former being that which lies in the plane of (yz). Then it will be found that

$$p^2 - \dfrac{b^2c^2\left(\dfrac{Y^2}{b^4}+\dfrac{Z^2}{c^4}\right)\left(\dfrac{Y^2}{b^2}+\dfrac{Z^2}{c^2}-1\right)}{\left(\dfrac{Y^2}{b^2}+\dfrac{Z^2}{c^2}\right)^2},$$

$$q^2 = a^2 - \dfrac{a^2}{\dfrac{Y^2}{b^2}+\dfrac{Z^2}{c^2}}.$$

From these two equations we can deduce

$$Y^2 - \frac{a^2b^2}{b^2-c^2}\,\frac{b^2-\dfrac{a^2p^2}{q^2}}{a^2-q^2},$$

$$Z^2 - - \frac{a^2c^2}{b^2-c^2}\,\frac{c^2-\dfrac{a^2p^2}{q^2}}{a^2-q^2}$$

Then substituting the values of Y^2 and Z^2 in the equation which they have to satisfy, we obtain finally

$$\frac{a^4}{b^2 c^2}\left(\frac{b^2}{h^2}+\frac{c^2}{k^2}\right) = \frac{q^4}{q^2-p^2}.$$

This shews that q^2 must be greater than p^2, and if we put

$$p^2 = q^2\,(1-e^2)$$

we obtain a *constant* value of $\dfrac{q^2}{e^2}$; so that the directrices are in planes at a constant distance from that of (yz).

460. $(x'-x)\cos\theta - (y'-y)\sin\theta + \dfrac{c}{a}(z'-z) = 0,$ normal plane;

$(x'-x)\cos\theta - (y'-y)\sin\theta - \dfrac{a}{c}(z'-z) = 0,$ osculating plane;

required condition $c-a$.

CAMBRIDGE: PRINTED BY C. J. CLAY, M.A. AT THE UNIVERSITY PRESS.